T0190099

Water Resources Development and Management

More information about this series at http://www.springer.com/series/7009

Mrinmoy Majumder
Editor

Application of Geographical Information Systems and Soft Computation Techniques in Water and Water Based Renewable Energy Problems

 Springer

Editor
Mrinmoy Majumder
Hydro-Informatics Engineering
National Institute of Technology
Agartala, Tripura
India

ISSN 1614-810X ISSN 2198-316X (electronic)
Water Resources Development and Management
ISBN 978-981-13-4825-9 ISBN 978-981-10-6205-6 (eBook)
https://doi.org/10.1007/978-981-10-6205-6

This Springer imprint is published by Springer Nature
The registered company is Springer Nature Singapore Pte Ltd.
The registered company address is: 152 Beach Road, #21-01/04 Gateway East, Singapore 189721, Singapore

Preface

The present monograph tries to highlight some of the novelties in research and development in mitigation of problems faced by the water and water-based energy industry. The chapters published in the monograph depict innovative application of neural networks, multicriteria decision-making, geographical information systems in solving the different contemporary problems dealt under the area of water and water-based energy.

The solution provides an opportunity and methodology for the provision of technical, political and economic benefit not only for the environment, ecology but also for the socio-economic benevolence of the stakeholders.

Presently, the impact of urbanization and climate change has greatly disturbed the ecological and environmental equilibrium of many countries. Due to the uncontrolled extraction of natural resources and destruction of environment for sustenance of the need and luxury of the population which is benefitted by the technological advancements that is taking place both globally and regionally have induced global warming worldwide which in turn have created change in the regular pattern of climate.

The anomalies like excess sedimentation in surface water bodies, increase in frequency of extreme events, adversities in quality of water and above all these implications have greatly influenced the economic return from not only the water resources but also the energy extracted from such resources.

The water-based energy resources such as hydropower, wave energy must be managed optimally for maximum utilization and return on investment.

The manuscripts published in this dissertation focus on the predicament as well as provide novel solutions with latest technologies to mitigate intricacy of the quandary so that pinnacle return can be possible even under extreme uncertainty.

As for example in the first chapter, Poon and Hwee try to estimate the inflow in reservoirs considering the sedimentation problems of the Teriang Reservoir in Malaysia where as in Chap. 2, the Suryanarayana et al. try to entice the effect of extreme events during the twentieth century in Surat district of India.

In Chap. 3, the quality of water of irrigation waters was analysed by Roy, through a new and alternative water quality index which was developed by the application of fuzzy logic decision-making (FLDM) method. This type of decision-making procedures was found to be optimal when the importance of two parameters is nearly equal.

The economic connotation of hydro power plant under climatic extremities was examined in the Chap. 4 where novel approaches like cascading of multicriteria decision-making (MCDM) with neuro-genetic modelling systems were used to estimate the financial impact of a small scale hydro power plant in a peri-urban state of Tripura.

In the fifth and sixth chapters, the location selection, which is the major factor for optimal utilization of available wave energy potential against minimum investment, was tried with the help of MCDMs like analytical hierarchy process (AHP) and polynomial neural networks (PNN) by Chakraborty and Ghosh.

In the last two chapters, the cost incurred in hydro power plants for augmenting utilization of potential was minimized by the application of various novel optimization tools; hybridization of wind and water power was practiced, and innovative solution for optimal scheduling was attempted for maximum utilization of energy by Mishra et al. and Dhillon et al., respectively.

The present monograph tried to accentuate the different solution available for reduction of uncertainty in pursuit of water and water-based energy resources which might be extremely beneficial for finding solutions for the problems imposed by uncertainties in the climatic pattern, large scale of urbanization and uncontrolled extraction of natural resources.

Agartala, India   Mrinmoy Majumder

Acknowledgements

The editor will like to acknowledge the contribution of all the authors and reviewers who have submitted and reviewed the manuscript for publication of this monograph. In the articles selected for this monograph, many data and information collected from different sources were cited. The editor on behalf of the authors will also like to tender his gratitude towards the publishers and owners of these sources and reports for granting permission to cite or reproduce the information.

The editor will take this opportunity to show his heartiest gratefulness towards the Series Editor, Publishing Editor and Project Coordinator for their encouragement, cooperation and support which have helped in great extent in completion of this monograph.

Last but not the least, the author and editor will like to thank their family, colleague and all those friends for their assistance and provision during the time of preparation of this manuscript.

Above all, the editor will consider this piece of literature a success if the would-be readers and reviewers of this book can really accelerate their endeavours for their objective of research.

Mrinmoy Majumder

Contents

Introduction

In recent years, the demand for energy has accelerated due to the burgeoning population and advent in technical proficiency. In 2008, total global energy consumption was 132,000 terawatt-hours (TWh) or 474 exajoules (EJ) whereas the average global power demand was 15 terawatts (TW). In 2012, mandate for energy amplified to 158,000 TWh (567 EJ) or an average power use of 18.0 TW.

The global power demand is satisfied by the conversion of primary energy into utilizable form of power, and the same is supplied to the consumers as and when required. Oil followed by coal/peat/shale and nature gas sector was found to supply about 81.10% of the global energy supply.

In 2012, world primary energy supply amounted to 155,505 TWh or 13,371 Mtoe, (IEA 2014) while the world final energy consumption was 104,426 TWh or about 32% less than the total supply. World final energy consumption includes products as lubricants, asphalt and petrochemicals which have chemical energy content but are not used as fuel. This non-energy use amounted to 9404 TWh (809 Mtoe) in 2012.

By the end of 2014, the total installed global power generating capacity is nearly 6.142 TW (million MW) which does not include the DG sets not connected to local electricity grids. In 2014, world energy consumption for electricity generation was coal 40.8%, natural gas 21.6%, nuclear 10.6%, hydro 16.4%, "others" (solar, wind, geothermal, biomass, etc.) 6.3% and oil 4.3%. Coal and natural gas were the most popular energy fuels for generating electricity. The world's electricity consumption was 18,608 TWh in 2012. This figure is about 18% smaller than the generated electricity, due to grid losses, storage losses and self-consumption from power plants (gross generation).

This evolution is the result of two contrasting trends: energy consumption growth remained vigorous in several developing countries, specifically in Asia (+4%). Conversely, in OECD, consumption was severely cut by 4.7% in 2009 and was thus almost down to its 2000 levels. In North America, Europe and the CIS, consumptions shrank by 4.5, 5 and 8.5%, respectively, due to the slowdown in economic activity. China became the world's largest energy consumer (18% of the total) since its consumption surged by 8% during 2009 (up from 4% in 2008).

Most of the world's high energy resources are from the conversion of the sun's rays to other energy forms after being incident upon the planet. Some of that energy has been preserved as fossil energy, some is directly or indirectly usable, for example, via solar PV/thermal, wind, hydro- or wave power. The total solar irradiance is measured by satellite to be roughly 1361 watts per square meter (see solar constant), though it fluctuates by about 6.9% during the year due to the Earth's varying distance from the sun. This value, after multiplication by the cross-sectional area intercepted by the Earth, is the total rate of solar energy received by the planet; about half, 89,000 TW, reaches the Earth's surface.

The estimates of remaining non-renewable worldwide energy resources vary, with the remaining fossil fuels totalling an estimated 0.4 yottajoule (YJ) or 4×1023 joules and the available nuclear fuel such as uranium exceeding 2.5 YJ. Fossil fuels range from 0.6 to 3 YJ if estimates of reserves of methane clathrates are accurate and become technically extractable. The total power flux from the sun intercepting the Earth is 5.5 YJ per year, though not all of this is available for human consumption. The IEA estimates for the world to meet global energy demand for the two decades from 2015 to 2035 will require investment of \$48 trillion and "credible policy frameworks".

According to IEA (2012), the goal of limiting warming to 2 °C is becoming more difficult and costly with each year that passes. If action is not taken before 2017, CO_2 emissions would be locked-in by energy infrastructure existing in 2017. Fossil fuels are dominant in the global energy mix, supported by \$523 billion subsidies in 2011, up almost 30% on 2010 and six times more than subsidies to renewables.

In the year 2000, the top three consumers of energy were the USA, China and Russia which consumed a total amount of 4071 Mtoe of energy, but in the year of 2015, the total consumption of energy rose to 6179 Mtoe, and in the list of top three consumers, the name of India get included and China becomes the highest consumer of energy (IEA, 2016).

The data as depicted above clearly indicate a steady increase in consumption of energy sources due to the burgeoning population, technological advances and economic stability achieved.

In plain language, R/P basically gives us the length of time the reserves would last if their usage continues at the current rate. Here are estimated world total R/P ratios for the main conventional fuels: oil—46 years, natural gas—58 years, coal—118 years. Of course, the usage is constantly changing and occasionally, new deposits are found. That's why the R/P ratios are corrected every year.

In contrast, renewable energy (RE) resources, as the name implies, are constantly replenished naturally and will never be exhausted. Their use generally has a much lower environmental impact than that of conventional fuels.

All these factors, coupled with the government incentives and mandates, result in growing interest in using alternative sources of energy. While some green technologies are large scale, many of them are also suited to private homes, especially in rural areas.

Alternative energy resources like power extracted from solar, hydro, wind or many other renewable forms of energy are analysed so that a suitable alternative, which can either substitute or assist the conventional fuel resources in compensation of the rising demand for energy, can be identified.

While raw forms of energy are both free and practically infinite, the equipment and materials needed to collect, process and transport the energy to the users are neither one. Currently, the RE costs are generally higher than that of fossil-based and nuclear energy. In addition to this, unlike well-established conventional designs, the advancement in different RE technologies still requires substantial investments. The economists often use so-called levelized energy costs (LEC) when comparing different technologies.

The LEC represents the total cost to build and operate a new power plant over its life divided to equal annual payments and amortized over expected annual electricity generation. It reflects all the costs including initial capital, return on investment, continuous operation, fuel and maintenance, as well as the time required to build a plant and its expected lifetime.

The complexity and cost involved in conversion of the renewables into utilizable energy forms is the major obstacle in the wide-scale application and consequent substitution of the conventional energy resources.

This table compares the US average levelized electricity cost in dollars per kilowatt-hour for both non-renewable and alternative fuels in new power plants, based on US EIA statistics and analysis from Annual Energy Outlook 2014. Note, that the numbers for each source are given for a different capacity factor, which complicates direct comparison. Notwithstanding, I believe these figures are useful in comparing different power generation methods.

Power plant type	Cost $/kW-hr
Coal	$0.095–0.15
Natural gas	$0.07–0.14
Nuclear	$0.095
Wind	$0.07–0.20
Solar PV	$0.125
Solar thermal	$0.24
Geothermal	$0.05
Biomass	$0.10
Hydro	$0.08

The availability of the renewable resources is the other concern which have prevented wide-scale application of the alternative forms.

Our dashboards present data on what's known as the levelized cost of energy. In essence, this analysis offers an apples-to-apples comparison of the costs of financing, building, operating and maintaining a power plant. The values are expressed in dollars per megawatt-hour.

One of the most widely used levelized cost studies is conducted by Lazard, an international financial advisory and asset management firm. Their latest version of the study, version 8, was released in late 2014. The graphic below summarizes the cost components of 16 different energy technologies evaluated by Lazard: 10 of them are alternative (which includes mainly low-carbon, renewable technologies), and six are conventional (which includes fossil fuel sources and nuclear).

Onshore wind has the lowest average levelized cost in this analysis at $59 per megawatt-hour, and utility-scale photovoltaic plants weren't far behind at $79. By comparison, the lowest cost conventional technologies were gas combined cycle technologies, averaging $74 per megawatt-hour and coal plants, averaging $109. These numbers are the average of Lazard's low- and high-end estimates (see their study for more about their cost calculations).

Looking across the 16 technology types, the 10 alternative technologies cost an average $147 per megawatt-hour, $18 less than the conventional approaches. "Certain Alternative Energy generation technologies", by dividing the costs among capital, fuel, and operations and maintenance (O&M), you can see some dramatic differences among the technologies. Many renewable technologies, such as wind, solar and geothermal, may not be cheap to build, but they have no fuel costs once they're up and running, and generally have lower O&M costs as well.

The levelized cost of some wind and solar technologies has plummeted in recent years. The graphic below shows that the average cost of onshore wind has fallen from $135 per megawatt-hour in 2009 to $59 in 2014. That's a 56% drop in five years. The cost of utility-scale photovoltaic technology has plunged from $359 per megawatt-hour in 2009 to $79 in 2014, a 78% decline. Lazard attributes these falling costs to "material declines in the pricing of system components (e.g. panels, inverters, racking, turbines), and dramatic improvements in efficiency, among other factors" [3].

In the USA, renewable energy for electric power, transportation, industrial, residential and commercial purposes is the fastest-growing energy source, increasing 52% from 2000 to 2013 from 6.1 to 9.3 quadrillion British Thermal Units (Btus).

In 2013, renewable energy was responsible for nearly 10% of total US energy consumption (all sectors), up from a little more than 6% in 2005, due to strong growth in biomass (including biofuels) and wind power.

In 2013, renewable energy was responsible for 12.9% of net US electricity generation with hydroelectric generation contributing 6.6% and wind generation responsible for 4.1% of this total.

Globally, renewable energy was responsible for approximately 22.1% of electricity generation with hydro generation accounting for 16.4% of the total in 2013.

The U.S. Energy Information Agency projects that solar power will be the fastest-growing source of renewable energy in the USA with annual growth averaging 7.5% in the period from 2012 to 2040. In 2013, solar generation accounted for 1.8% of total renewable generation. In 2040, this is projected to climb to 10%.

In 2013, renewable ethanol and biodiesel transportation fuels made up 21.5% of total US renewable energy consumption, up from just 12% in 2006.

However, the energy extracted from flowing water was found to be most inexpensive and readily available which according to the World Bank has the maximum potential to replace or support contribution of fossil fuels in mitigation of the rising energy demand.

Water power captures the energy of flowing water in rivers, streams and waves to generate electricity. Conventional hydropower plants can be built in rivers with no water storage (known as "run-of-the-river" units) or in conjunction with reservoirs that store water, which can be used on an as-needed basis. As water travels downstream, it is channelled down through a pipe or other intake structure in a dam (penstock). The flowing water turns the blades of a turbine, generating electricity in the powerhouse, located at the base of the dam.

Large conventional hydropower projects currently provide most of renewable electric power generation. With 1000 gigawatts (GW) of global capacity, hydropower produced an estimated 3750 TWh of total global electricity in 2013. Note that in 2013, total global electricity generation was 22,865 TWh. Hydropower operational costs are relatively low, and it generates little to no greenhouse gas emissions. The main environmental impact is to local ecosystems and habitats; a dam to create a reservoir or divert water to a hydropower plant changes the ecosystem and physical characteristic of the river.

The USA is the fourth-largest producer of hydropower after China, Canada and Brazil. In 2011, a much wetter than average year in the US North-west, the USA generated 7.9% of its total electricity from hydropower. The quantity of electricity generated each year depends on the amount of precipitation that falls over a area.

Small hydropower, generally less than 10 megawatts (MW), and micro-hydropower (less than 1 MW) are less expensive to develop and have a lower environmental impact than large conventional hydropower projects. In 2011, the total amount of small hydro installed worldwide was 106 GW—China had the largest share at 55.3%, followed by India at 9% and the USA at 6.9%. Many countries have renewable energy targets that include the development of small hydro projects. In the USA, the Federal Energy Regulatory Commission (FERC) approved more than 50 project permits in 2009.

Hydrokinetic electric power, including wave and tidal power, is a form of unconventional hydropower that captures energy from waves or currents and does not require dam construction. These technologies are in various stages of research, development and deployment. In 2011, a 254 MW tidal power plant in South Korea began operation, doubling the global capacity to 527 MW.

Low-head hydro is a commercially available source of hydrokinetic electric power that has been used in farming areas for more than 100 years. Generally, the capacity of these devices is small, ranging from 1 kW to 250 kW.

Pumped storage hydropower plants use inexpensive electricity (typically overnight during periods of low demand) to pump water from a lower-lying storage reservoir to a storage reservoir located above the powerhouse for later use during periods of peak electricity demand. Since this technology uses more electricity than it generates, it is not considered to be renewable energy. Note that it is economical to do this since the revenues that a generator receives during times of peak electricity generation far exceed the costs that they pay to pump the water during times of low electricity demand.

World energy consumption is likely to grow 56% to 820 quadrillion Btus from 2010 to 2040 with most of this development coming from emerging countries. Renewables are projected to be the fastest-growing source of energy with consumption of hydroelectricity and other renewables set to increase from 11% of total energy consumption in 2010 to 15% in 2040 [2].

Hydropower is a renewable, efficient and reliable source of energy that does not directly emit greenhouse gases or other air pollutants, and that can be scheduled to produce power as needed, depending on water availability.

There are about 78,000 megawatts of hydropower generation capacity in the USA.

Over the last decade, depending on water availability, hydroelectricity provided 5.8 to 7.2% of the electricity generated in the USA and averaged more than 70% of the electricity generated annually from all renewable sources, although this share is falling as the renewable capacity from other sources grows.

More than half of the total US hydroelectric capacity for electricity generation is concentrated in three western states–Washington, California and Oregon–with approximately 26% in Washington alone. Canada is a major electricity supplier to New York, New England, the Upper Midwest, the Pacific North-west and California.

Hydropower could offer at least 80,000 megawatts of additional generation capacity. Small, micro and low hydropower are developing hydro technologies, while the efficiency and capacity of existing hydroelectric generators can be improved. Only about 3% of the roughly 80,000 dams in the USA have hydropower plants and can generate electricity.

Existing hydropower is very inexpensive to operate (generation costs 2–4 cents per kilowatt-hour). The levelized cost of electricity of new hydropower projects, of less than 50 megawatts, (6–14 cents per kilowatt-hour) and incremental hydropower projects (adding generating capacity to existing dams; 1–10 cents per kilowatt-hour) puts them among the least expensive forms of low-carbon electricity.

The effects of climate change on water availability are expected to affect hydropower generation.

The mandate and accessibility of global water and energy resources highlight the necessity of recognizing the trade-off between the demand and availability of this two resources. However, if the demand for energy can also be satisfied from water availability after mitigating the demand of water from the consumers, then present uncertainties can be reduced and as power from water is completely free of carbon compounds, and both terrestrial and aquatic environment can be protected and conserved.

In numerous places of the world, hydropower was counted in as a part of watershed management programmes. Wind energy is found to be the fastest-growing renewable energy technology for producing grid-connected power among different non-conventional energy sources. Apparently, a total volume of 17,353 MW wind power has been documented up to 31 March 2012 in the country, which is about 70% of the cumulative deployment of the grid-interactive renewable power. The figures of identified potential are suggestive of the identified precisely feasible potential only and not of the potential that is techno-economically viable. Interim revised estimated by C-WET is 100,000 MW at 80 m. height [1].

In the present group of articles, the trade-off and the optimization that can be achieved by integrating power from water in water resource management programmes were highlighted. Various solutions and tools were applied to optimally utilize these two natural resources. Although the integration of the two resources were not shown directly, the basic goal of all the papers was to manage water or water-based energy resources by implementing soft computational tools in such a manner that both water and energy can be conserved and utilized for the mitigation of demand from the ever-growing base of consumers.

References

1. Center for Climate and Energy Solution. (2017). *Renewable energy*. Retrieved from https://www.c2es.org/energy/source/renewables on 6th June 2017.
2. Energy Innovation. *Comparing the costs of renewable and conventional energy sources*. Retrieved on June 6, 2017 from http://energyinnovation.org/2015/02/07/levelized-cost-of-energy/
3. Rozenblat, L. (2017). *Why is it important to use renewable energy?, your guide to renewable energy*. Retrieved on June 6, 2017 from http://www.renewable-energysources.com/
4. IEA. (2014). *2014 Key World Energy Statistics*. pp. 6–38, Retrieved from http://www.iea.org/publications/freepublications/.

Abstract

This book highlights the application of geographical information system (GIS) and nature-based algorithms to solve the problems of water and water-based renewable energy resources. The irregularity in availability of resources and inefficiency in utilization of the available resources have reduced the potentiality of water and water-based renewable energy resources. In the recent years, various soft computation methods (SCM) along with GIS were adopted to solve critical problems. The present book collects various studies where many SCM were used along with GIS to provide a solution for optimal utilization of the natural resources for satisfying the basic needs of the population as well as fulfilling their burgeoning energy demand. The articles depict innovative application of soft computation techniques to identify the root cause and to mitigate the uncertainty for optimal utilization of the available water resources. The advantage of SCM and GIS was used to maximize the utilization of water resources under cost and time constraints in face of climatic abnormalities and effect of rapid urbanization. The case studies were divided into two parts: water-based problems and water-based renewable energy problems. Part I deals with solving the problems of the "resource", and Part II includes the studies which maximize the efficiency of the resource utilization process.

Keywords Soft computation methods · Geographical information systems · Water resource management · Water-based energy resource management

Part I
Water Based Problems

Chapter 1
Review of Reservoir Sediment Inflow Estimation for Teriang Reservoir, Malaysia

Ching Poon Hii and Hock Hwee Heng

Abstract The purpose of this review is to summarize the current standard of practices for estimating reservoir sediment inflows for the entire duration of their useful service life, normally 100-year ARI design period. To achieve this objective, a review of the past and contemporary literatures and conventions on sedimentation issues in Malaysia as well as other regional countries is imperative. A case study was carried out using the proposed Teriang reservoir. (CA = 59 km^2) that is located mainly in the forested headwater region of Sg. Teriang which eventually joins Sg. Pahang further downstream in the state of Pahang, Malaysia. Both the low and high (Type I and II respectively) curves are adopted for predicting the suspended load concentration (mg/l) using normalized flow discharge per unit area (Q/km^2; m^3/s/km^2). Alternative method by coupling the sediment rating and flow duration curve (both daily and monthly) are also used for checking purpose. The limitations and insights of total sediment loadings estimates for the case study are discussed in order of importance.

Keywords Reservoir sediment inflow · Low and high (type I and II) curves · Sediment rating and flow duration curve

1.1 Introduction

The purpose of this review is to summarize the current standard of practices for estimating reservoir sediment inflows for the entire duration of their useful service life, normally 100-year ARI design period. To achieve this objective, a review of

C.P. Hii (✉)
Department of Civil Engineering, Faculty of Engineering,
Universiti Malaysia Sarawak, Kota Samarahan, Malaysia
e-mail: Hiicp6588@yahoo.com

H.H. Heng
Faculty of Engineering and Science, Universiti Tunku Abdul Rahman,
Sungai Long, Malaysia

© Springer Nature Singapore Pte Ltd. 2018   3
M. Majumder (ed.), *Application of Geographical Information Systems and Soft Computation Techniques in Water and Water Based Renewable Energy Problems*, Water Resources Development and Management, https://doi.org/10.1007/978-981-10-6205-6_1

the past and contemporary literatures and conventions on sedimentation issues in Malaysia as well as other regional countries is imperative.

The secondary objective is to estimate the sediment loading into the proposed Teriang reservoir. (CA = 59 km^2) that is located mainly in the forested headwater region of Sg. Teriang which eventually joins Sg. Pahang further downstream in the state of Pahang. In the past practices of SMHB (and its predecessor, Binnie and Partners), estimates of reservoir sediment loading are primarily based on its two major regional water resources studies, i.e. Pahang and Johor water resources master plans [20, 21]. The conclusion in both studies recommended **260–300** tonnes/km^2/year (or 260–300 m^3/km^2/year if the bulk sediment density is assumed as unity) of total sediment load per unit catchment be allocated for planning purposes for forested catchment.

This is mainly inferred due to limitations and inadequacies of observed sediment gauging records throughout Malaysia, notwithstanding hardly any stations are located at the individual and specific project sites of interest. It is indisputable uncertainties are prevailing in estimating the reservoir sedimentation and the collection of sediment data not only has to deal with the temporal pattern of the sediment but also the spatial distribution of the sediment sources as well. The suspended sediment gauging program is extremely costly and time consuming compared to other categories of hydrometric sampling program. At most one spot sample is collected during periodic streamflow gauging process vis-à-vis a continuous stage recording. In addition, the other vital component that added to the suspended sediment loads, bed load measurement is virtually nonexistent in the official JPS hydrometric sampling endeavor (except a few pioneers in universities are starting to sample bed load in their respective research endeavors).

Owing to these shortcomings in obtaining reliable and long-term observed sediment records, inferences from regional studies are imperative and seemed to be a viable option. By doing so, with limited sediment records, a range of sedimentation rate, i.e. 260–300 tonnes/km^2/year for forested catchment is therefore recommended upon consulting with the past studies carried out by SMHB and its predecessor.

The ranges of recommended loadings are correlated well with the some short-term field observation, i.e. Terengganu basin study [18], Pergau dam feasibility study [17], and NK/SMHB (2000). Besides the reasonableness of the proposed range is also suited for reservoir catchment of predominantly upland forested land use, except for a few others, such as Ringlet and three hydroelectric reservoir schemes on the main stem of Sg. Perak (Temengor, Chenderoh, Kenering). Due to their unique locations in the middle reaches of Sg. Perak and changing land use pattern (primarily logging and large scale commercial farming in the highland) within the catchment over time, they have suffered to variable degrees of soil erosion and sedimentation problems.

The existing land use upstream of the proposed Teriang dam site (CA = 59 km^2) is predominantly forested with moderately undulating terrain. Although there are some minor small-scale sugarcane plantations in the vicinity of dam site, this will stop when the dam is constructed and no human settlement or development activity in the upper reaches of Sg. Teriang. Accurate quantification of sediment inflows into the reservoir is required so that an appropriate dead storage allowance (and subsequently as a basis for setting the minimum operating level [MOL]) could be made. Normally, the dead storage allowance of 100-year ARI is sufficed for major reservoir projects in Malaysia.

1.2   Past Studies on Sedimentation Issues in Malaysia

Past studies on sedimentation were concerned with a primary objective of allocating reservoir dead storage by estimating long-term sediment inflows into the reservoir. The period of consideration is conventionally assumed the same as the service life of the reservoir, i.e. 100-year ARI. Major dam/reservoir design studies dedicated mostly a brief subsection on sedimentation in light of scarcity of information available to mandate in depth and detailed undertakings. Therefore, due to both scarcity and unavailability of observed concurrent flow and sediment records and in addition the fact that most of the dam/reservoir projects that were usually situated in the upper catchments where logistic constraints prevent the implementation of hydrometric sampling program, recourses are therefore made to infer sediment estimates from gauged records nearby as well as in the region.

Unfortunately, as mentioned earlier, most of these suspended sediment gauging stations are located at downstream of the basin. Bed load sampling program is not carried out. Nationwide and regional studies are also conspicuously absent in Malaysia due to perhaps inadequate sediment records for in depth and meaningful analyses. Extensive records nevertheless are not in existent in Malaysia. However statewide water resources master plan studies for the states of Pahang and Johor (PWRS and JWRS, [20, 21]) have include a short and concise chapter exclusively on sedimentation issues in reservoirs and preliminary estimates of sediment inflows were made for each potential reservoir scheme.

Other than these, individual or standalone feasibility and subsequent detailed design studies on reservoir scheme throughout Malaysia also carried out a substantial amount of works on sedimentation issues. These recent major undertaking were Kenyir dam feasibility study [18] and Pergau hydroelectric project feasibility study [17], Gerugu dam feasibility study [10], Babagon dam [11], Murun Hydroeletric project [9], Bakun dam [8], NK/SMHB (2000) to mention a few.

For most past studies and designs undertaken by SMHB (PWRS 1992; JWRS 1994), some **260–300** tonne/km^2/year of total sediment load has being adopted. It is reasonable and adequate as most of these reservoirs are located in the headwaters and remain pristinely forested.

Evidently, low sediment accumulation was vividly observed in the Durian Tunggal reservoir in Melaka when it dried up in the early 1990s (this is the only known case where a reservoir was drawn to its bottom in Malaysia). Unfortunately this anecdotal observation was not in any way systematically quantified and reported in writing.

Other pertinent recent study was the Interstate Water Transfer Project from Pahang to Selengor (NK/SMHB 2000) where an alternative methodology of sediment load estimations were carried out based on selected regional records and estimates in both states of Pahang and Selangor.

Another sources of information of sedimentation are the research undertakings by local institutions and/or collaborative program with foreign organizations. Some of these projects such as in Sabah and Sarawak are mainly concerned with the impact of forest harvesting or logging in a watershed. Extensive sampling program is a prerequisite for assessing and quantifying the impact of catchment disturbances on water resources. One of these on going programs is Danum Valley Field Centre (DVFC; [6]).

Estimating reservoir sediment inflows is inherently depend on the long-term (say more than 10 years; or as proposed by experts in sedimentation engineering, at least half of the design life span of the reservoir project) background or baseline sampling of the sediment in a watershed. This is nevertheless lacking in Malaysia and most of the research endeavors carried out are more related to short-term impact of land use changes in watershed, i.e. conversion to agricultural plantation or forest harvesting.

Another category of sediment and erosion engineering practices in Malaysia is to estimate the sediment yields due to hill-slope land development. The pervasive problems of mud and debris avalanches mandate estimates of sediment loads in response to heavy downpours and other weather phenomena in relatively high rainfall intensity humid tropical climatic region such as Malaysia. The sediment yields due to either natural or man-made erosions are calculated based on Universal Soil Loss Equation (USLE; attributed Wischmeier and Smith 1965), soil loss = RKLSCP) and a cohort of its variants from time to time, i.e. Modified (MUSLE), Revised (RUSLE) and reports on application to other countries outside USA, CALSITE and DUSLE, essentially European based models. These hill-slope process procedures though developed mainly on agricultural plots by Agricultural Research Services (ARS) of U.S. Department of Agriculture (USDA) are commonly used for mainly agricultural soil loss computation and they have also found receptive audiences in Malaysia, especially in the Environmental Impact Assessment (EIA) communities.

The earlier reporting on sediment yield using these types of simple equations (an aggregate or cumulative effects of coefficients) apparently did not take into consideration the applicability of extrapolating relatively small field plot results to fairly larger land areas. The other shortcomings, such as suitability of adopting similar coefficients (five of them) developed in temperate climatic region such as the Midwestern and northern region of USA (calibrated against more than 10,000 plot-years of data from erosion plots at 42 experiment stations located in 23 states in the United States), such as rainfall erosivity (R), soil erodibility (K), topographic (L and S), and crop management and soil conservation (C and P) factors etc. in Malaysia are absent in Malaysia. In addition the cardinal rules of calibration and validation processes are also inadvertently absent perhaps due to lack of or unavailability of observed records.

This, however does not in any way preclude the researches carried out subsequently on assessment of soil erosion and sedimentation on hill slope development [23]. Most of these issues could also be found in abundance in most of legally mandated EIAs carried out on large-scale development projects in Malaysia. Over the years with enormous experiences and knowledge learned on soil erosions. Some of these USLE coefficients are developed locally to suit condition in tropic climates of Malaysia. References on this specific issue of sedimentation and erosion should be made to Tew and Faizal [23] and Chan [5].

Review on hydrology and sedimentation in the tropical forest of South East Asia is also reported in Douglas [6] and Douglas et al. [7]. Some observed measurements and results of monitoring program could be found in these research articles.

1.3   Sg. Terengganu Basin Study: Feasibility of Hydroelectric Project [18]

This feasibility study encompassed programs for potential hydroelectric projects identification in the upper Sg. Terengganu basin. Several potential sites were investigated. Preliminary estimate of sediment load was required for determining the minimum operating level of the reservoir scheme.

For this Study, historical measurements of flow and sediment discharges were lacking, only a limited gauging (based 200 samples during November 1974 to January 1975) carried out ad hoc in the Sg. Terengganu basin indicated some 96 m^3/km^2/year of total sediment load into the Kenyir reservoir (CA = 2600 km^2). The flow duration sediment rating method included bed load equal 20% of that suspended load and a bulk density of 1.0 tonnes/m^3.

1.4   Silting Study at Pedas Impounding Dam [3]

The Pedas dam was commissioned in 1932. It is located on one of the tributaries of Sg. Pedas, Sg. Beringin, at the upper catchment of the greater Sg. Linggi basin. It drains a small catchment area upstream of the dam of about 5.65 km^2. The gross storage of the reservoir is about 45.5 million liter. However range line survey carried out in 1978 indicated the storage was reduced by silitation to about half of its original capacity. A net loss of 23 million liter of the storage reservoir over a period of 47 years is equivalent to about 23,000 m^3 per 47 year, 489 m^3/year. In terms of per unit area, the estimated siltation is about 87 m^3/km^2/year, which was considered reasonable for upland watershed with predominant forest land-use in Malaysia. Desilting was proposed in this earlier study to restore its gross storage and hence the reliable yield for water supply as its original purpose intended.

1.5 Water Resources Development for East Negeri Sembilan, Melaka, Northeast Johor [19]

This was part of the comprehensive water resources assessment carried out in the mid of 1980s, which scarcity of sedimentation information was available for meaningful analysis of reservoir sediment allocation. As a result, inference on sedimentation/soil loss rate was made to the earlier AUSTEC (1974) study in Sg. Pahang basin.

The AUSTEC study (1974) proposed a sediment/soil-loss rate of 100 tonnes/ km^2/year for unlogged forest. For conservatism, however higher rate was adopted in this study and a bulk sediment density of 1.5 tonnes/m^3 was also assumed. Therefore in terms of volume, the sedimentation rate was 67 m^3/km^2/year. A factor of 5.0 was adopted for logged forest and for rubber estate grown on a 30-year rotation, a much higher factor, i.e. 7.50 were adopted. This was to account for human interference over natural sedimentation in the basin.

With these assumed values in mind, the sedimentation rate for 100-year life time for unlogged and other forms of disturbed land use were computed. Table below shows the sedimentation accumulation for Upper Muar (now known as Talang), Gemenceh, and Juaseh reservoirs.

Reservoir sedimentation accumulation for various reservoirs

Reservoir	CA (km^2)	100-year sediment accumulation (MCM)
Upper Muar (Talang)	148[a]	0.80–4.00
Gemencheh	37	0.60–1.10
Juaseh	29	0.15–0.70

[a]Residual area = 111 km^2 after deducting Kelinchi dam CA, 37 km^2

1.6 Pergau Dam Feasibility Study [17]

In the Pergau Hydroelectric Project, the estimates of suspended sediment load were determined using the flow-duration sediment rating curve method. The long-term daily flow duration curve of Kuala Yong dam site (CA = 86 km^2) was used. Total sediment load was finally obtained by assuming bed load was 10% of suspended load and a bulk density of 1.50 tonnes/m^3. Other than these, adjustment was also made on using shorter time interval, i.e. less than 1 h on the sediment estimate. Basing on the past experiences of SMEC in Ok Tedi region projects in Papua New Guinea, the sediment load was 4.0–8.9 times higher when determined using a 30-min time interval vis-à-vis 1 h. Similar observations were also evidenced in some catchments in Java, Indonesia, the ratios between 30-min and 1-h time interval varied from 1.30 to 2.80. To adapt to the local condition, estimates of sediment load using both hourly and daily time internal flow duration curve was made and the ratio was 1.29. Subsequently, a round-off figure of 1.30 was used for adjustment.

With a range of suspended load estimates using various flow duration curves. The most realistic range of estimates of the annual sediment inflows to Kuala Yong reservoir was considered to be 72,000 to 178,000 m^3/year (or 837–2070 m^3/km^2/year and 558–1380 tonnes/km^2/year). The corresponding denudation rates are about 0.80–2.0 mm/year. Due to uncertainty in the sediment estimates (i.e. errors in instrumentation and relatively short records), conservatism is therefore exercised by adopting the highest sediment inflows 178,000 m^3/year. The estimates were comparatively higher than other any adopted sediment loads for reservoir project in Malaysia.

1.7   Pahang Water Resources Study [20]

The purpose of this specific segment in the hydrological subsection of Pahang water resources study was to estimate the rate of sedimentation in the proposed reservoirs so that adequate dead storages are allocated for each scheme. To do so, estimates of the annual sediment yields of the catchment in Pahang were required.

A total of nine (9) concurrent flow and suspended sediment JPS gauging stations in Pahang were chosen for further analysis. The suspended sediment loads range as low as 38 tonnes/km^2/year to 203 tonnes/km^2/year. With the exception of Sg. Lepar @ Jambatan Gelugor gauging station, where the observed suspended sediment load was 1033 tonnes/km^2/year. All these gauging stations are in the low land and middle reaches of the greater Sg. Pahang, except Sg. Lipis @ Benta and Sg. Bentong @ Kuala Marong, so their representations of dam site catchment located mostly in the headwaters is doubtful. On the other hand, Sg. Bentong @ Kuala Marong can be deemed representative of dam site catchments since it is located in the higher altitude but it drains and traverses a substantial part of the Bentong town catchment. This may render itself invalid of representing mostly forested land use prevailing in most reservoir schemes in Malaysia. Table below shows the suspended sediment loading at respective JPS gauging stations.

Observed sediment load in Sg. Pahang system

Catchment	Catchment area (km^2)	Period of measurement	Average annual suspended sediment load		
			Tonnes	Tonnes/km^2	m^3/km^2
Sg. Semantan @ Jambatan Keretapi	2920	1972–1979	181,000	62	41
Sg. Jelai @ Jeram Bungor	7320	1972–1989	914,000	125	83
Sg. Jelai @ Kuala Medang	2630	1973–1988	300,000	114	76
Sg. Bentong @ Kuala Marong	241	1970–1988	20,000	83	55

(continued)

(continued)

Catchment	Catchment area (km^2)	Period of measurement	Average annual suspended sediment load		
			Tonnes	Tonnes/km^2	m^3/km^2
Sg. Pahang @ Termeloh	19,000	1963–1989 65; 85–87 missing	3,470,000	183	22
Sg. Teriang @ Jambatan Keretapi	2000	1972–1988	76,500	38	26
Sg. Kuatan @ Bukit Kenau	582	1975–1988	177,000	304	203
Sg. Lipis @ Benta	1680	1965–1988	225,000	134	89
Sg. Lepar @ Jambatan Gelugor	560	1972–1988	579,000	1033	689

Excerpted from PWRS (1992); Bulk density = 1.5 tonnes/m^3

Other than JPS stations, another sixteen (16) streamflow stations maintained and operated by TNB (operated until the early 1990s) were located in the upper northwestern catchment of Sg. Pahang. These stations were conveniently located at the potential TNB dam sites for hydropower projects. Out of these, six (6) concurrent streamflow and sediment gauging stations were available for analysis. The rating curves of these combined JPS and TNB (15) stations were subsequently produced and attached as appendices to the report.

Review of recent sedimentation studies were also carried out on sedimentation within Sg. Klang basin [2]. Suspended sediment load ranging from 165 to 2283 tonnes/km^2/year were obtained from thirteen gauging stations in the Sg. Klang basin. On average, the sediment load was 803 tonnes/km^2/year.

A further attempt was also made to classify the sediment load attributed to prevailing land use upstream of the gauging stations. The lowest sediment yield was recorded, i.e. 165 tonnes/km^2/year for entirely forested catchment in upper Sg. Gombak, a major northwestern tributary of Sg. Klang basin. On the other hand, moderately forested Sg. Ampang (CA = 88 km^2) with some land development activities at the downstream end gave some 379 tonnes/km^2/year. The data collected was only up to late 1980s. It is anticipated the sediment load would be higher with much more intensive land development in the past 20 years.

As expected the highest load i.e. 2283 tonnes/km^2/year was found in urbanized catchment and with large percentage of former mining lands in the sandy Sg. Jinjang basin. In between, other different land uses with rapid land developments such as housing that produced some middle ranges of sediment yield, from 1265 to 1759 tonnes/km^2/year. These documented sediment yield though qualitatively suggested a positive relationship between sediment loading and land use utilization in the basin. Of interest to the reservoir management was the sediment load of these forested catchments, Sg. Gombak and Sg. Ampang. This reinforced the fact that entirely forested catchment basically produces less than 200 tonnes/km^2/year of suspended sediment yield. By including says, another bed load (assume 25% of the suspended load) to it, the total sediment load is about 225 tonnes/km^2/year

Not many studies (especially monitoring of regular sediment inflows) on reservoir sedimentation were carried out in Malaysia. The notable and significant study was the tracking of historical occurrence of sedimentation in the Ringlet reservoir (or known as Sultan Abu Bakar dam; CA = 183 km^2) in Pahang/Perak border. Range line surveys were carried out since 1965 till 1991. During these periods, some 11 bathymetric surveys were carried out. By summing up the difference between bed profiles of the reservoir between two time periods, sediment inflow could be confidently inferred. The earlier periods from 1965 to mid 1980s did not show any significant sedimentation occurring in the reservoir. Only after mid 1980s, the inferred sedimentation rates increased several fold to 1814 m^3/km^2/year. Reasons attributed to the increases of sedimentation rate are land use clearing for cash crops farming and large-scale land development in the upper catchment especially in Cameron Highland.

Other range line survey in Klang Gates reservoir (CA = 77 km^2) was undertaken in a comprehensive United States Bureau of Reclamation (USBR) Sg. Klang flood mitigation program. The capacity of Klang Gates reservoir was resurveyed in 1976. The sediment accumulation for a period of 17 years since its commission in 1959 estimated a total of 750,000 m^3 of sediment had accumulated in the reservoir. This implied on average, a sediment yield of 573 m^3/km^2/year or 859 tonnes/km^2/ year if the bulk density of the sediment was 1.50 tonnes/m^3.

Finally provisional recommendation for preliminary dam design was made, i.e. 260 tonnes/km^2/year on the sedimentation rate for various reservoir sites in this study. The rationale of adopting this value was not known when the observed sedimentation rates in the existing Klang Gates (CA = 77 km^2) reservoir was obviously known to be higher, i.e. 859 tonnes/km^2/year though with shorter period of records. An exception case was reservoir sites that are mostly affected by anticipated future development, such as in the Sg. Bentong basin, a higher sedimentation rate of 500 tonnes/km^2/year was otherwise provided. Recommendation was also made for regular review on sedimentation issues if the potential schemes were to be taken up for implementation.

In practice, the allocation of dead storage may not be a critical issue to be reckoned with in the first place since the minimum operation levels (top of the sediment dead storage zone [MOL]) and the siting of draw-off facilities are mostly determined by the topographical considerations and constraints. As such, the dead storage zone in all cases is likely to exceed the necessary allowance for sedimentation.

1.8 Johor Water Resources Study [21]

The segment on sedimentation issues in this Study was basically derived from the earlier Pahang water resources study [20]. The allocation of dead storage for potential reservoir projects was mostly based on evidences presented in the earlier 1992 report in the neighboring catchment. Another four (4) more stations with concurrent streamflow and suspended sediment gauging program were also included. The sediment yield ranges from 53 tonnes/km^2/year in the Sg. Bekok catchment to 161 tonnes/km^2/year in the Sg. Johor catchment over a limited sampling period from 1980 to 1988. Table below summarizes records from these four gauging stations.

Observed sediment load in the State of Johor

Catchment	Catchment area (km^2)	Period of measurement	Average annual suspended sediment load		
			Tonnes	Tonnes/km^2	m^3/km^2
Sg. Bekok	350	1980–1988	18,526	53	35
Sg. Muar @ Buluh Kasap	3130	1978–1988	212,556	68	45
Sg. Johor @ Rantau Panjang	1130	1981–1988	714,469	161	107
Sg. Linggiu	209	1983–1988	12,666	61	41

Excerpted from JWRS (1994); Bulk density = 1.5 tonnes/m^3

In the previous Pahang water resources study [20], an allowance of 260 tonnes/km^2/year over 100 year was recommended for forested or cultivated with tree crops. However much higher rates, i.e. 500 tonnes/km^2/year were adopted for catchments with high erosion potential occurrence due to developments in catchment areas.

1.9   Detailed EIA of Bakun Hydroelectric Project [8]

A literature review was carried out for most of the sedimentation studies in both Peninsular and Borneo states of Sabah and Sarawak. The review focused primarily on the impact of logging on sediment yield in primarily tropical rain forest catchments. Most of the studies in Malaysia as reported were concentrated on the effect of land use modifications on experimental hill slope plots and for catchment sizes up to 140 km^2. As recognized by the fact that the land use modification is one of the most important and dominant variables on sediment yields, the forest-covered basins were generally yielding relatively less sediment vis-à-vis other land uses via human intervention. The most pertinent abstract from this review was the quantitative documentation of sediment and erosion yields. As quoted from other relevant studies, the sediment yield for small experimental field plots covering undisturbed rain forest range from less than 100 tonnes/km^2/year to just over 300 tonnes/km^2/year. This however, was not sufficient for accurately assessing the sediment inflows into the proposed Bakun reservoir with its sizable catchment area of 14,750 km^2. Therefore, modelling studies on sedimentation were carried out instead.

An earlier study on Bakun hydroelectric project by SAMA (1983) reported the total average annual sediment inflow into the Bakun reservoir at 9.0 million tonnes/year or 610 tonnes/km^2/year. This was amounting to about 0.61 mm/year denudation rate if conservatively 1.0 tonnes/m^3 bulk sediment density is assumed.

This estimation included the bed load component at an assumed percentage, i.e. 20% of the suspended sediment loadings. This was a very preliminary estimate of reservoir sediment loading based only on limited field observation data. For brevity, both components of the total load are summarized in table below.

Annual sediment load estimate

Type	Annual load	
	Total (million tonne)	Specific (tonne/km^2)
Suspended load	7.5	508
Bed load	1.5	102
Total	9.0	610

Bakun dam catchment CA = 14,750 km^2

The sediment deposit in the Bakun reservoir was then computed for 50- and 100-year accumulation. Table below show the respective deposition volumes of the Bakun reservoir.

Sedimentation volume and capacity

Period (year)	Sediment volume (million m^3)	Reduction in capacity (%)	Sediment level @ Dam (m a.s.l.)
50	380	0.90	77.0
100	740	1.70	78.9

Excerpt in entirety from SAMA (1983)

This follow-up study was carried out subsequently after SAMA (1983) with additional observed suspended sediment data at Sg. Linau and Bakun Rapids in the Bakun catchmemt.

This reassessment predicted the average suspended sediment yield to be 6.4 million tonnes/year or 432 tonnes/km^2/year for a long-term (1983–1998 and 1999–2043) simulation run period using modeling approach (by coupling flow and sediment mass loading). This suspended load estimate was slightly lower than the previous estimate by SAMA (1983), i.e. 7.5 million tonnes/year or 508 tonnes/km^2/year. This study argued that the present estimate was more reliable than the pre-1983 estimate in SAMA (1983) as using a more recent flow data measured in the vicinity of the proposed Bakun dam site.

The bed load is however estimated using Schocklitsch equation (1934) as measurement was not available in the basin. As expected, with high uncertainty in most of the bed load sediment transport equations, the predicted bed load (1983–1998 and 1999–1943) was calculated at 7.2 million tonnes/year or 488 tonnes/km^2/year. This was amounting to about 116% over of the suspended sedimente yield. For information, a much lower value, i.e. 20% of the suspended sediment yield was assumed in the SAMA (1983) report.

The scenario run of mass loading is made possible by availability of long-term flow time series using LASCAM and accurate discharge-concentration rating curves. Six (6) catchment management scenarios were carried out to determine the impact on the sediment yield due to various extents and degrees of catchment disturbances, i.e. commercial logging and residual biomass clearing in the Bakun basin. The scenario runs ranged from optimum to worst, were considered and were compared with the base scenario, SAMA (1983). The total average sediment load based on simulation results of the management scenario runs (1, 3, 5, and baseline) is summarized in table below.

Average annual total sediment load (1983–1998 and 1999–2043)

Scenario	Suspended sediment load (million tonne/year)/ (tonnee/km^2/year)	Bedload sediment load (million tonne/year)/ (tonnee/km^2/year)	Total sediment load (million tonne/year)/ (tonnee/km^2/year)
1 "worst"	29.8 (2020)	7.5 (508)	37.3 (2528)
3 "most likely"	24.1 (1634)	7.4 (502)	31.5 (2136)
5 "best"	18.8 (1275)	7.3 (495)	26.1 (1770)
6 Baseline	6.4 (434)	7.2 (488)	13.6 (922)
SAMA (1983)	7.5 (508)	1.5 (102)	9.0 (610)

() denotes unit of tonnes/km^2/year; Catchment Area @ dam site = 14,750 km^2

Compared to SAMA (1983), the annual average total sediment yield estimated in EIA report [8] is higher than the former due to significantly high bed load estimation in latter. The suspended load component of both studies is nevertheless comparable at 6.4 and 7.5 million tonnes/year for SAMA (1983) and Ekran [8] respectively.

Catchment disturbances due to both continued logging and somewhat "forest-unfriendly" harvesting techniques after post impoundment period, is depicted in scenario 1 where the total sediment load is the highest among others.

If some conservative and environmental-friendly techniques are implemented such as "helicopter logging", some degree of mitigation and management of the residual biomass in the impoundment, and strict control on the issuance of the timber harvesting licenses as described in scenario 3, the total sediment load, mainly suspended load into the reservoir would be reduced by about 16%.

Further reduction (of about 17%) on the total sediment load into the reservoir is possible albeit "near-ideal" scenario 5 where strictly no logging and harvesting or any other types of catchment disturbances are allowed in the Bakun dam basin. This "best" scenario is most unlikely to be achieved given present circumstances.

In summary, the "most likely scenario" is best described by scenario 3 where the total sediment load i.e. 2136 tonnes/km^2/year, slightly triples the previous estimate, 610 tonnes/km^2/year (SAMA 1983).

Sedimentation volume and capacity

Period (year)	Sediment volume (million m^3)	Reduction in capacity (%)	Sediment level @ dam (m a.s.l.)
50	380	0.90	77.0
100	740	1.70	78.9

Excerpt in entirety from SAMA (1983)

It was therefore inferred by linear proportion that if the sediment load was increased from the baseline scenario to scenario 5 ("most likely"), the accumulated sediment volume would increase to about 4.5 times the estimate used in the design study by SAMA (1983). The 50- and 100-year sediment accumulation volume was therefore revised to 4.1 and 7.7% of the gross storage capacity respectively.

1.10 Interstate Raw Water Transfer: Pahang-Selangor (NK/SMHB 2000)

Estimation of reservoir sediment loads was required for the determination of minimum operating level (MOL) for proposed Kelau (CA = 331 km^2) and Telemong (CA = 360 km^2) reservoirs in the western region of greater Sg. Pahang basin. Due to scarcity of observed sediment records in the basin and surrounding, recourse was made to infer the sediment inflows using a regionalized approach by constructing a relationship between the sediment concentration and discharge or flow per unit area for concurrent streamflow and sediment gauging records of JPS hydrometric stations in both Kelau and Telemong basins.

Two regionalized curves, namely Type I and II were derived based on observed flow and sediment records furbished by JPS (see Figs. 1.1 and 1.2). Type I curves comprised mostly records of river basins from the state of Selangor. These stations are Sg. Langat @ Kajang (CA = 389 km^2), Sg. Semenyih @ Rinching (CA = 225 km^2), Sg. Gombak @ Jalan Tun Razak (CA = 122 km^2), Sg. Langat @ Dengkil (CA = 1240 km^2) and Sg. Selangor @ Rasa (CA = 321 km^2). These stations are more or less representative of urbanized and developed basins that result in higher suspended sediment concentrations in the river.

On the other hand, Type II basins mainly comprise of rural setting land use though with sparse plantations and most important of all, the upper catchments remained forested virgin jungle. These representative gauged records are Sg. Lepar @ Jambatan Gelugor (CA = 560 km^2), Sg. Bernam @ Jambatan SKC (CA = 1090 km^2), Sg. Jelai @ Kuala Medang (CA = 2630 km^2), Sg. Lipis @ Benta (CA = 1670 km2), Sg. Bentong @ Kuala Marong (CA = 241 km^2), and Sg. Selangor @ Rantau Panjang (CA = 1405 km^2). The curves are manually enveloped in order to give the best upper bound representative curves. In a step further, they are fitted into linear and power based equations for ease of computation later as shown in Figs. 1.1 and 1.2.

LOW CURVE or TYPE II

Q/km2 (m3/s/km2)	ORIGINAL CONC (mg/)l	CALCULATED CONC (mg/l)
0.0058	10	6
0.021	100	121
0.066	500	463
0.100	700	734
0.200	1100	1028
1.000	2200	2245

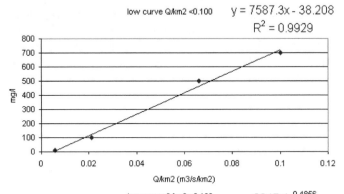

low curve Q/km2 <0.100 $y = 7587.3x - 38.208$

$R^2 = 0.9929$

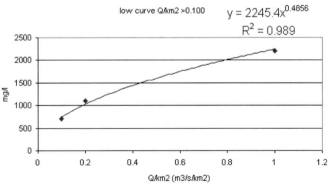

low curve Q/km2 >0.100 $y = 2245.4x^{0.4856}$

$R^2 = 0.989$

Comparison of best fit and extracted data from NK/SMHB 2000

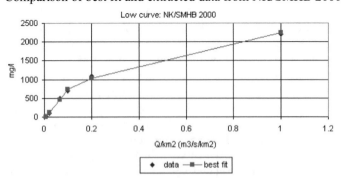

Fig. 1.1 Type II sediment unit flow curve

HIGH CURVE or TYPE I

Q/km2 (m3/s/km2)	ORIGINAL CONC (mg/)l	CALCULATED CONC (mg/l)
0.002	27	15
0.005	90	95
0.009	190	202
0.02	500	495
0.097	2000	1760
0.5	6300	5943
1	9000	9942

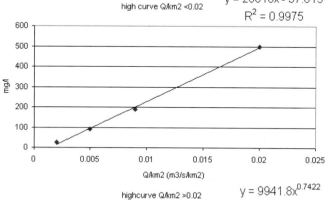

high curve Q/km2 <0.02

$$y = 26618x - 37.815$$
$$R^2 = 0.9975$$

highcurve Q/km2 >0.02

$$y = 9941.8x^{0.7422}$$
$$R^2 = 0.9927$$

Comparison of best fit and extracted data from NK/SMHB 2000

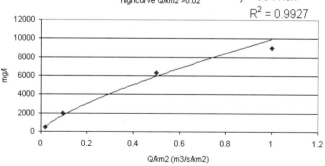

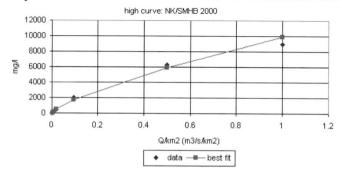

high curve: NK/SMHB 2000

Fig. 1.2 Type I sediment unit flow curve

For purpose of quantifying the total sediment inflows into both Kelau ($CA = 331$ km^2) and Telemong ($CA = 360$ km^2) reservoirs, both Type I and II curves are used to give a range of both higher and lower sediment loadings using daily reservoir inflow/discharge time series derived earlier in the hydrological assignment. Bearing in mind that, the curves that are derived from suspended load measurement and therefore do not include bed load measurement, alternative and indirect inference of the bed load is imperative. Handicapped by virtually the fact no bed load measurement has been carried out in Malaysia (except reported by Lai [13], for this study, bed load is assumed 20% of the suspended load and is added to the suspended load accordingly. This is a common practice to circumvent the lacks of proper measurement on bed load transport (PWRS and JWRS 1992, 1994; [17]). Only in the Bakun EIA study [8], estimate of bed load component of the reservoir sediment load was based on bed load sediment transport formula, in this case, bed shear type of equation attributed to Schocklitsch [1]. Utility of sediment transport formulae in sediment estimate is rather rare in Malaysia. This is perhaps the pioneer case adopted in Malaysia.

In addition, a ratio/coefficient adjustment on effect of different time scales, i.e. daily versus hourly is also included in the computation of total sediment load. The adjustment coefficient is assumed 1.20, which is more or less consistent with previous SMEC/SMHB [17] study in Pergau Hydroelectric project. To convert sediment mass to volume, a bulk density of 1.30 tonnes/m^3 is assumed. The annual total sediment loads expressed in terms of mass and volume per unit catchment area are shown in table below.

	Kelau reservoir CA = 331 km^2		Telemong reservoir CA = 360 km^2	
	Low type II	High type I	Low type II	High type I
Volume per annum (m^3/year)	79,000	11,500	99,000	126,000
Volume per unit area per annum (m^3/km^2/year)	239	347	275	350
Mass per unit are per annum (tonnes/km^2/year)	311	451	356	455

It would be unreasonable to adopt high or Type I (high) curve for most of the upland reservoir that is predominantly forested and known to export less sediments vis-à-vis other disturbed land use type. Comparing with PWRS and JWRS (1992, 1994), the total sediment loads estimated for both Kelau and Telemong reservoirs seemed reasonable and in the same order of magnitude. For most of the estimates discussed previously in this report, it is assumed that the catchment land use remains the same as the time when the sediment load is estimated. Any changes in land use, mostly of human intervention such as large-scale logging operations will definitely contribute to increase of sediment inflows and thus accelerate in filling the dead storage allocated for their entire service life.

1.11   Gerugu Dam Design [10]

Provision of sediment load allocation in the proposed Gerugu reservoir (CA = 13.6 km^2) is the upper limit of the recommended per unit area total sediment load, i.e. <u>300 m^3/km^2/year</u> (assumed sediment bulk density of 1.0 tonnes/m^3). This was obviously deduced from the earlier PWRS (1992) study. With assumption of 92% trap efficiency and unity sediment bulk density, the loss of storage after 100 year was 0.38 MCM. The corresponding denudation rate was 0.30 mm/year. The MOL was preset at 11.0 m msl and the dead storage for 100 year was only slightly above 10.0 m msl.

1.12   Murum Hydroelectric Project Feasibility Study [9]

Sg. Murum is one of the major tributaries of Sg. Balui at the upper catchment of Greater Sg. Rajang basin. It is located in the headwater region of the current Bakun HEP project. The proposed dam site during the phase I of the feasibility study drains some 2750 km^2. The corresponding reservoir water surface is approximately 200 km^2 at a Maximum Operating Level (MOL) of 531 m msl. The corresponding gross storage at MOL is about 7000 MCM. No detail sedimentation study was carried out in phase I but promised of detailed and a more close up study during phase II stage. Inferences were therefore made from regional studies, provision of denudation rate of 0.50–1.0 mm/year (1.38 and 2.76 MCM/year or 500 to 1000 m^3/km^2/year respectively) was proposed. Even for allowance for 100 year of constant erosion rate, the reservoir will only silt up with about 69 to 138 MCM of the dead storage. Obviously, the amount of accumulated sediment constituted only a small fraction of the gross reservoir storage of about 7000 MCM.

1.13   Kelinchi Reservoir [22]

A sedimentation rate of 100 tonnes/km^2/year for unlogged forest was proposed based on the Pahang Water Resources Study. Provision of higher rate, a factor of 5.0 for logged forest. For a catchment of 38 km^2, the 100-year sediment inflow into the reservoir was less than 1 MCM, which was insignificant compared to the dead storage allocated when setting the MOL.

1.14   Babagon Dam Design Report [20]

The reservoir is located in the upper catchment of Sg. Moyog basin, a relatively small catchment south of Kota Kinabalu. The direct supply reservoir drains some 30 km^2 of catchment area and a live storage of about 17 MCM. Of interest to this study, was the assumed sediment load of 240 tonnes/km^2/year for dead storage allocation. The value was deduced from other reports in the Peninsula, such as, Pahang and Terengganu river basin studies, Johor river projects, etc. The sediment loads reported by these studies ranged from as low as 93 tonnes/km^2/year to as high as 330 tonnes/km^2/year. For not so obvious reason, it seemed that a value somewhere in between the both end of the extreme was adopted. The 100-year estimated load for a catchment area of 30 km^2 was about 0.60 MCM, assuming full sediment trap efficiency.

This study highlighted the estimation of sediment load mostly under average condition should be cautioned. On the other hand, up to 75% of the annual sediment may occur during a single storm event. However no appropriate study been cited.

1.15   Reports by Researchers on Sedimentation

In Malaysia, some important water resources multidisciplinary approache type of researches were also been carried out from time to time by local institutions, such as Forest Research Institute Malaysia (FRIM) and Universities (notably, Department of Forest Science of University Putra Malaysia [UPM] and Universiti Sain Malaysia [USM] mainly on sedimentation in urban catch basins). Sedimentation studies are an integral part of these research endeavors. Valuable information on observed sedimentation records could be excerpted from these respective studies. Most important of all, due to their relative smaller sizes, monitoring program could be confidently planned and thus accurate concurrent sediment and flow measurements could be obtained. Another aspect of the elaborated studies that are pertinent to the quantification of sedimentation in river basin is the comparative studies of the impact of deforestation or afforestation on the sediment yield. These studies are mostly carried out by experts in the forestry science, such as researchers from UPM. One of the recent studies was monitoring and measuring the suspended load resulting from forest harvesting in the Sg. Weng Experimental watershed in Kedah ([ICUH] [12]).

Other than the above, some collaborative programs are also being initiated on forest hydrological and ecological network research. Danum Valley Field Center (DVFC) is one of such programs in association with the Universiti Malaysia Sabah (UMS), Forest Research Institute of Malaysia (FRIM), Sabah Forest Department, FACE Foundation (Netherlands), New England Power (USA), and the Royal Society of London [6].

1.16   Sg. Weng Experimental Watershed [12]

In this recent study partially funded by Muda Agricultural Development Authority (MADA), suspended loads were measured for resulting change due to forest harvesting in Sg. Weng experimental watershed. Long term monitoring program is on going and the recent publication only reported a few years of observation, i.e. from 1997 to 2000. The experimental watersheds are basically small catchments ranging from 2.8 to 8.4 km^2 located in the south of the Sg. Muda reservoir catchment. Four (4) compartmental plots were demarcated within Sg. Weng basin. The compartments represent four types of land use control from controlled and undisturbed, two methods of logging practices, and monitoring of downstream of the Sg. Weng basin for effect of land use modification.

The controlled basin or undisturbed basins basically produced less sediment than the others of variable degrees of treatment and logging practices. The controlled basin, as designated as W1 measured suspended load varied from 97 to 134 tonnes/km^2/year. For the same period, the basin (designated as W3) that underwent conventional logging practices using crawler tractor-winch lorry system produced significant suspended load ranging from 169 to 5487 tonnes/km^2/year. The other basin W2 using conventional practices but with some best management practices incorporated in the logging program such as wider 30 m stream buffer strip and 40° slope limit demonstrated a much less suspended sediment production, ranging from 118 to 206 tonnes/km^2/year. The contrast though with short sampling period was evident convincingly by comparing both three types of land uses disturbances.

In his earlier paper [13], observed average suspended sediment yields of 54–90 tonnes/km^2/year were documented for undisturbed (Sg. Lawing, CA = 5.0 km^2) and much larger but partly disturbed (Sg. Lui CA = 68 km^2; 20% logged in the late 1970s) basin in Selangor respectively. These perhaps could be considered representative of forested catchment in the central region of Peninsular Malaysia. The other documentation on the sediment yield in a mechanized logging basins, i.e. Sg. Batangsi (CA = 20 km^2) and Sg. Chongkak (CA = 13 km^2) that indicated average suspended sediment yield of 2826–2458 tonnes/km^2/year after first year of operation. However decreased load was observed after second year of logging in Sg. Chongkak. The reduction was almost half of the earlier observed suspended sediment yield, i.e. 1335 tonnes/km^2/year.

Other than suspended sediment gauging program carried out in the Lai [13] study, bed load sampling program was also carried out albeit at a much smaller scale due to high costs and logistical constraints. Bed loads of 125 and 22 tonnes/km^2/year were observed at Sg. Lawing (CA = 5 km^2) and partly disturbed Sg. Lui (CA = 68 km^2) catchment. These accounted for some 230 and 25% of the concurrent suspended load measurement. Correspondingly for other catchments, Sg. Batangsi (CA = 20 km^2) and Sg. Chongkak (CA = 13 km^2), the bed load measured amounting to 1264 and 619 tonnes/km^2/year first year after logging. These again represented some 45–25% of the suspended load. Second year after logging in Sg. Chongkak, the bedload was reduced to 334 tonnes/km^2/year or 25% of the suspended load measured during the same period. Overall Lai [13] suggested that bed load may form a large part of the total sediment yield i.e. 20–70% and varied amongst monitored sites undertaken in the sampling program.

1.17   TNB Seminar [15, 16]

Realizing the impacts and consequences of sedimentation and erosion on the hydroelectric projects under its charge, TNB has identified its reservoir scheme which were affected by the sedimentation problems. Of many schemes, Ringlet, Chenderoh, and Kenering scheme were selected for further evaluation. To adequately quantify the sediment inflows into the reservoirs, range line surveys were carried out. Amongst them, the Ringlet reservoir was the most affected by sedimentation problem due to massive land use clearing for agricultural purpose.

1.17.1   Chenderoh Reservoir (CA = 6560 km^2; Storage: Live = 84.4 MCM; Dead = 9.06 MCM)

The Chenderoh reservoir/dam was the first of a series of four (4) cascade hydropower dams built along Sg. Perak. It drains some 6560 km^2 of the upper basin. However about 85% of its inflows are regulated by three other upstream schemes, namely, Kenering (1984), Bersia (1983), and Temengor (1978). The Chenderoh reservoir was commissioned about 75 years ago. Other dam schemes were gradually added on to the greater Sg. Perak scheme. With the completion of Kenering scheme in 1984, it is fully regulated by three (3) upstream reservoirs. It was therefore anticipated the sediment inflows would then be proportionately decreased with much smaller water flows into the reservoir. Although range line surveys were carried out as early as in 1954, unfortunately the systematic compilation of the sediment accumulation had not being made until 1988.

The depletion rate (presumably of the dead storage) of 0.68% per year since 1954 was reported. Between 1960 and 1962, a slightly smaller rate was observed, i.e. 0.61%. The impoundment of the second reservoir at the most upper catchment, Temengor reservoir, was expected to reduce the sediment load substantially. A constant conservative rate of 0.68% was assumed until 1976, and followed by a lower rate, i.e. 0.50% for the remaining period, from 1977 till present. It was evident that the sedimentation rate was within the designated life span of the reservoir, presumably, a design period of 100 years. Lower rate was also anticipated with the completion of Kenering reservoir in 1984.

Assessment on Kenering reservoir was not carried out as it was only being operated for a short period of about 8 years (up to 1992 at the time of this study). With its sizable gross reservoir storage, 352 MCM, it was reckoned that the sedimentation should not be a critical issue at least for the time being. It was indicated that the sediment loads were contributed mostly by the Sg. Rui catchment, which was about 20% of the size of the residual catchment area downstream of Bersia dam to Kenering dam.

1.17.2   Ringlet Reservoir (CA = 183 km²; Storage: Live = 4.7 MCM; Dead = 1.6 MCM)

A short description was also presented on the sedimentation of Ringlet reservoir (CA = 183 km²; 111 and 72 km² respectively from Sg. Telom and Sg. Bertam catchment) in Cameron Highland. The sedimentation problem at the Ringlet is mostly due to land development for agricultural projects in the upper catchment in the early 1980s. The economic development of the highland into eco-agricultural and tourism centers was accelerated and more previously forested land was turned into open space plot. By doing so, these human activities caused extensive and serious erosion and degradation. The sediments were mostly flushed downstream in massive quantity and eventually reached the Ringlet reservoir downstream. As a result, measures, such as desilting operation by draglines and occasional flushing during high flows have to be undertaken. In light of the problem, reservoir hydrographic survey was initiated in order to quantify the amount of sediments being flushed into the reservoir. Starting 1965 till 1992, the sedimentation rate of the reservoir was meticulously documented. As indicated, the sedimentation problems aggravated from as low as 52 m³/km²/year to some 1814 m³/km²/year in the earlier 1990s at the time of this report. The loss of capacity was estimated at 31% since the beginning of operation till early 1990s (31% as calculated without desilting program from 1981 onward). Both structural (mechanical dredging and flushing) as well as non-structural (land use planning and control) measures were one way or another taken to address these pressing problems. This program was timely implemented else the reservoir life will be ended with sediments prematurely.

As summarized in the report earlier, the Ringlet reservoir that was originally designed to cater for a cumulative of sediment deposit (dead storage 1.60 MCM) within a span of approximately 80 years (1.60 MCM per 80 year equivalent) is silted up especially since 1980s coincidental with the massive agricultural expansion program in the watershed. The sedimentation rate as tabulated and documented by TNB indicated from 1965 to 1990, the total sediment inflows to the Ringlet reservoir were 7016 m³/km² for 26 years. This translated into some 1.28 MCM, which was 80% of the dead storage. With this accelerated inflows, the dead storage allocated for the Ringlet reservoir will filled up within 30 years vis-à-vis the designed 80 years. With the implementation (starting from 1981) of the frequent dredging program, the dead storage occupied by sediment was only 31% in 1992. In other words, some 0.73 MCM (almost half of the allocated dead storage) of sediments had been dredged.

Another hydropower reservoir scheme, Batang Padang (CA = 210 km²) in the lower catchment, constructed during the same periods however does not suffer from the same fate as most of the upland catchment remains intact since its construction in 1968. The catchment area is located within the Tapah hills Forest Reserve, much of its natural state remains so that at the time of this report, 1992 sedimentation is not a pressing problem like the Ringlet. This vividly shows the influence and impact of human activities within the dam catchment on reservoir sedimentation.

1.18   Other Studies

At the time of this report, unfortunately the review on research undertaking by FRIM is not forthcoming. Information and findings could be of helpful in assessing the sedimentation rates of mainly forested catchment and the impact of land use modification by logging on water and sediment yield.

Other monitoring studies on the impact of logging and timber harvesting have been carried out in the Borneo state of Sabah and Sarawak under collaborative program with institutions such as Royal Society etc. Some of the medium-term results have been reported and published in scientific and scholastic journals, seminars, symposiums, colloquiums etc. as reported in Ekran [8], Douglas [6], Douglas et al. [7]. This secondary information were briefly mentioned in the literature review segment of the comprehensive EIA study. Specifically for this Study, individual papers and detailed findings are not forthcoming and therefore will not further review in this Study.

1.19   Sediment Records Availability in the Sg. Teriang Basin

There is no concurrent sediment sampling carried out at both Sg. Teriang @ Juntai (3,022,431; CA = 904 km^2) and @ Chenor (2,920,432; CA = 228 km^2) streamflow stations, but some short observed records are available further down stream near the railway bridge crossing at Sg. Teriang @ Jambatan Keretapi streamflow station (average annual suspended sediment load reported was 38 tonnes/km^2/year based on periods of record from 1972 to 1988.

Other than this, the nearest sediment sampling stations in the adjacent catchments are Sg. Linggi @ Jambatan Persekutuan and Sg. Muar @ Batu 57 Jalan Gemas-Rompin in the neighboring basins. Due to the lack of observed sediment records, inference of sediment discharges (both suspended and bed loads) could also be attempted as appropriate from recent regional studies carried out earlier such as SMHB [20, 21]. The latter two studies provide useful checks on the validity of the sediment estimation exercise.

As there are no sediment records in the form of flow discharge and sediment rating in the vicinity of the project area, an appropriate sediment rating curve or a regionalized enveloping curve should be selected for the purpose of sediment loading computation. Intuitively, the Sg. Teriang @ Jalan Keretapi station with some 17-year records could be chosen. However its sizable drainage area makes it less likely to be representative of the upper forested catchment area at Teriang dam site. It is well established that the sediment eroded in the upland could be deposited or aggradated before reaching downstream outlet. A famous plot of sediment delivery ratios (ratio of sediment eroded and observed at outlet of interest downstream) is inversely proportionate to the drainage areas [1]. Therefore this station is not selected for further analysis in this Study.

1.20 Methodology and Approach

Both the low and high (Type I and II respectively) curves are adopted for predicting the suspended load concentration (mg/l) using normalized flow discharge per unit area (Q/km^2; m^3/s/km^2). After obtaining the respective concentrations, sediment mass loading can then be computed accordingly. In this Study, two series of flow records are used. They are as follows:

1. 1948–1975 with 2 missing years (26 years record are available); and
2. 1948–2001 with synthetically generated runoff from 1976 onward using rainfall station at Jelebu (54 years record are available).

In addition, the following assumptions are made:

1. Monthly time series are used for computation, it is assumed that no adjustment will be made for taking into different time step of daily and monthly interval;
2. A coefficient of 1.20 is applied to the sediment load for reason of the difference of time interval adopted (hourly vis-à-vis daily) in the calculation. If no adjustment being made, it may underestimate the suspended sediment loading using daily time step and missing the critical intra-day variability during a storm event;
3. 20% is assumed for bed load unless it could be estimated from prevailing sediment transport equations. This is inline with previous studies [20, 21] and Lai [13]; and
4. 1.30 tonness/m^3 bulk sediment density is used for converting sediment mass to volume. If conservatism is desired, then a much lower density, i.e. 1.00 tonnes/m^3 could be adopted *in lieu*.

1.21 Sediment Rating Flow Duration Curve Method

Alternative method by coupling the sediment rating and flow duration curve (both daily and monthly) are also used for checking purpose [1, 14]. This methodology essentially uses both daily and monthly flow duration curve coupled with one or more sediment discharge rating curves to estimate the mass sediment loadings. This is similar to evaluating and quantifying the probability of occurrence of sediment inflows for a given time step, i.e. monthly or daily. The fractions of all possible probability of non-exceedance flow are extracted from the flow duration curve. As recommended by McCuen [14], some 20–25 segments should be extracted. These fractions (indirectly deemed as probability of occurrence) are used for adjusting the individual segmental sediment loading based on the flow discharge associated with specific fraction.

In this Study, suspended sediment loadings are calculated using both Type I and II (high and low) curves and flow duration curves. Adjustment for bed load fraction and the difference in time interval (hourly to daily) are then made accordingly as previously mentioned. The results are shown in Tables 1.2 and 1.3 for monthly duration and Tables 1.4 and 1.5 for daily duration respectively. For brevity the results are summarized in table below (Table 1.1).

Table 1.1 1948–1975; 1976–2001 synthetic records using Jelebu rainfall station co-relationship

YY? MM	Jan	Feb	Mar	Apr	May	June	July	Aug	Sep	Oct	Nov	Dec	Total
1948	74.38	58.18	73.58	69.67	79.15	48.16	43.93	29.11	23.82	60.85	81.73	60.34	702.90
1949	32.81	67.67	46.31	93.49	101.65	54.56	55.84	46.31	54.04	84.18	50.47	119.11	806.44
1950	63.53	60.24	69.08	114.75	96.34	66.10	43.66	42.08	42.51	48.69	80.17	61.95	789.09
1951	93.48	76.28	65.89	79.65	68.03	56.35	52.15	27.59	81.21	85.23	125.27	109.81	920.93
1952	91.07	69.41	80.89	44.06	58.66	45.36	28.66	30.00	56.76	36.43	110.94	71.51	723.74
1953	25.98	35.15	50.30	56.61	71.73	43.55	52.15	36.96	70.45	102.96	76.20	12.05	634.09
1954	105.07	59.27	69.64	104.25	88.12	60.70	89.73	49.76	44.32	94.23	102.20	188.18	1055.48
1955	135.79	81.53	103.23	96.06	94.23	89.14	90.53	99.53	27.22	31.61	88.13	77.30	1014.28
1956	61.42	39.86	34.93	51.48	71.46	46.63	40.50	53.73	55.57	88.39	83.51	123.88	751.35
1957	48.96	25.57	49.50	54.04	100.84	52.51	36.80	32.57	54.07	95.03	97.33	128.91	776.14
1958	84.18	56.66	72.80	48.73	81.53	36.62	203.83	46.07	14.07	95.81	95.02	21.16	856.48
1959	42.35	42.80	61.39	90.18	96.88	81.21	64.07	39.96	38.41	140.80	155.52	169.41	1022.97
1960	75.96	54.97	83.65	66.59	89.19	46.63	45.26	38.38	31.75	48.96	139.35	113.03	833.72
1961	77.81	49.74	39.96	59.18	68.30	32.27	29.65	43.93	36.11	30.96	40.23	61.95	570.08
1962	65.62	25.11	42.35	68.66	66.69	39.71	30.19	59.03	66.10	60.61	96.84	57.69	678.60
1963	36.51	32.73	25.39	26.65	51.93	29.96	23.81	13.77	22.55	40.23	83.51	53.73	440.77
1964	40.71	41.09	65.35	56.51	90.80	40.23	64.31	39.10	54.56	59.54	73.38	89.73	715.31
1965	46.34	39.19	37.31	31.10	68.03	25.14	30.80	58.12	60.39	81.42	136.86	114.37	729.08
1966	89.99	63.87	83.83	82.68	67.76	49.51	51.16	52.76	46.40	97.49	105.24	85.17	875.87
1967	94.08	107.12	81.49	88.30	69.69	64.31	53.48	34.15	53.78	78.91	132.78	94.06	952.16
1968	60.30	25.23	34.00	56.39	71.75	77.79	87.41	65.72	72.14	70.80	78.79	96.92	797.22
1969	77.94	52.98	54.37	59.36	79.55	51.32	43.12	44.73	55.21	101.78	90.46	83.57	794.39
1970	123.80	31.70	47.35	28.10	39.91	36.29	40.44	40.98	46.14	56.51	55.73	82.23	629.17
1971	141.42	53.46	80.62	40.95	43.66	44.32	64.82	72.05	56.51	52.23	59.88	116.24	826.16

(continued)

Table 1.1 (continued)

YY? MM	Jan	Feb	Mar	Apr	May	June	July	Aug	Sep	Oct	Nov	Dec	Total
1972	52.76	46.21	38.57	72.06	55.71	44.06	26.78	22.53	51.84	56.51	74.91	69.10	611.05
1973	39.10	25.64	37.50	38.36	34.02	37.84	24.11	28.95	29.42	48.24	101.09	72.69	516.96
1974	30.40	30.59	37.02	39.63	29.84	35.25	25.39	25.74	61.95	25.58	43.80	39.64	424.83
1975	61.07	35.56	66.69	48.47	35.09	41.73	58.98	45.69	52.51	43.87	75.82	49.10	614.58
1976	52.71	26.72	33.85	48.41	52.67	54.35	51.70	61.36	55.83	67.79	65.53	93.45	664.36
1977	81.01	73.46	43.83	37.56	62.21	63.35	52.97	45.37	37.94	88.69	80.70	68.35	735.44
1978	59.30	35.82	33.24	79.05	78.33	74.25	65.93	47.77	44.47	57.95	84.62	74.49	735.22
1979	51.49	51.50	46.45	90.18	69.11	63.04	51.83	41.65	65.10	58.99	84.39	63.28	737.01
1980	48.48	34.16	44.65	40.63	54.11	40.75	79.02	60.25	83.45	80.51	96.99	83.62	746.62
1981	54.33	44.05	53.54	84.45	88.68	59.70	35.78	23.31	63.85	47.68	72.88	74.30	702.54
1982	56.04	58.95	66.23	92.04	76.00	57.32	40.94	63.14	63.64	85.23	89.01	60.38	808.90
1983	55.88	39.13	38.87	42.97	37.04	40.90	47.41	54.36	64.09	54.98	88.17	57.95	621.72
1984	72.68	89.70	89.82	72.72	61.31	52.40	56.82	44.90	44.22	51.08	71.95	88.23	795.84
1985	70.78	54.91	56.83	50.09	69.28	41.54	64.53	41.48	68.18	56.69	84.17	96.52	755.01
1986	94.79	79.53	93.12	77.17	83.95	66.28	61.01	36.56	71.91	84.88	81.83	67.62	898.66
1987	47.94	30.60	41.16	82.79	78.57	72.96	39.30	56.21	77.44	95.32	82.46	90.35	795.09
1988	58.85	54.22	57.99	76.62	69.02	93.37	72.18	95.85	75.24	64.34	94.78	54.40	866.87
1989	68.33	30.12	61.26	68.27	63.77	43.61	55.11	48.80	73.85	65.80	72.79	46.65	698.38
1990	42.83	27.04	48.81	64.59	80.06	68.10	51.24	37.33	69.32	85.83	79.21	76.92	731.29
1991	51.35	50.52	51.99	66.23	94.86	88.77	67.00	47.12	36.96	71.13	65.34	99.84	791.10
1992	63.01	85.58	46.72	59.33	56.74	49.53	60.72	45.18	63.58	41.74	71.66	77.73	721.51
1993	60.32	49.17	52.95	88.39	86.04	80.89	70.12	57.66	71.50	73.11	127.19	117.57	934.92
1994	82.11	72.82	76.94	100.17	71.13	59.27	29.49	41.16	46.83	58.54	61.96	61.33	761.75
1995	76.69	53.49	77.55	55.96	59.28	58.28	44.49	69.60	56.21	79.76	84.82	104.34	820.47

(continued)

Table 1.1 (continued)

YY? MM	Jan	Feb	Mar	Apr	May	June	July	Aug	Sep	Oct	Nov	Dec	Total
1996	85.64	66.17	60.26	81.33	66.51	68.00	58.88	64.49	55.89	64.67	60.39	92.16	824.39
1997	60.17	83.98	67.97	123.55	73.12	66.86	39.37	51.18	53.81	80.45	73.81	94.24	868.52
1998	65.38	49.06	37.16	26.37	33.69	49.15	67.63	91.39	65.00	51.25	57.49	68.94	662.51
1999	65.99	49.57	63.82	56.33	72.76	63.22	47.68	46.83	78.46	88.54	99.35	89.75	822.29
2000	76.61	83.93	96.42	70.27	56.69	42.08	34.53	41.90	49.41	70.95	98.83	106.07	827.68
2001	97.88	62.96	77.10	80.10	65.66	52.03	33.74	31.69	47.30	66.79	69.54	83.27	768.06
MEAN	68.03	52.31	58.36	66.32	69.09	54.58	53.35	47.52	54.50	68.71	85.83	83.60	762.22
SD	24.47	19.53	18.84	22.70	17.97	15.75	26.69	17.10	15.80	22.11	23.63	31.02	127.85
SKEW	0.95	0.58	0.42	0.26	-0.32	0.58	3.49	1.05	-0.42	0.48	0.82	0.79	-0.25
From 1976 to 2001													
MEan	65.41	55.27	58.41	69.83	67.71	60.39	53.06	51.79	60.90	68.95	80.76	80.45	772.93
SD	14.54	19.37	18.03	21.68	14.44	14.21	13.33	16.27	12.92	14.47	15.12	17.81	74.83
SKEW	0.66	0.26	0.64	0.15	-0.47	0.56	0.01	1.16	-0.24	0.03	1.02	0.06	0.14
PWRS 1992 (1948–1967) (1969–1975)													
Mean	69.94	48.29	58.36	62.36	70.35	47.51	52.33	43.06	47.46	68.01	89.37	85.84	742.86
SD	31.89	16.06	19.55	23.88	21.73	14.53	35.85	17.18	15.99	28.66	28.73	41.19	164.47
SKEW	0.73	0.30	0.33	0.50	-0.35	1.32	3.25	1.39	-0.21	0.55	0.43	0.61	0.00

SG TERIANG @ JUNTAI CA = 904 km^2, unit = MCM/mth

Table 1.2 Sediment flow duration rating method: using low curve (Type I) and monthly flow duration curve

Low curve Frequency (%)	At juntai (MCM/mth)	At dam (Mld)	At dam site m³/s	At dam site %	Low curve Q (m³/s)/km²	Concentration (mg/l)	Concentration (kg/m³)	Flow m³/month	Mass loading kg/month	Mass loading tonn/month	adj by fraction
100	12.05	26	0.30	1.00	0.01	0	0.00	783,167	8.37	0	0.00
99	21.16	45	0.52	1.00	0.01	29	0.03	1,374,892	39,717	40	0.40
98	23.81	51	0.59	1.00	0.01	37	0.04	1,547,189	57,703	58	0.58
95	25.74	55	0.63	3.00	0.01	43	0.04	1,672,496	72,603	73	2.18
90	30.19	64	0.74	5.00	0.01	58	0.06	1,961,397	112,797	113	5.64
85	35.15	75	0.87	5.00	0.01	73	0.07	2,284,039	167,314	167	8.37
80	38.38	82	0.95	5.00	0.02	83	0.08	2,493,950	208,238	208	10.41
75	40.44	86	1.00	5.00	0.02	90	0.09	2,627,959	236,613	237	11.83
70	43.80	93	1.08	5.00	0.02	101	0.10	2,846,347	286,611	287	14.33
65	46.31	99	1.14	5.00	0.02	109	0.11	3,009,100	326,899	327	16.34
60	49.51	105	1.22	5.00	0.02	119	0.12	3,216,878	382,089	382	19.10
55	52.76	112	1.30	5.00	0.02	129	0.13	3,428,529	442,640	443	22.13
50	55.73	119	1.37	5.00	0.02	139	0.14	3,621,093	501,529	502	25.08
45	59.18	126	1.46	5.00	0.02	149	0.15	3,845,095	574,586	575	28.73
40	61.95	132	1.53	5.00	0.03	158	0.16	4,025,476	636,975	637	31.85
35	67.76	144	1.67	5.00	0.03	177	0.18	4,403,136	777,884	778	38.89
30	72.06	154	1.78	5.00	0.03	190	0.19	4,682,157	890,931	891	44.55
25	79.65	170	1.96	5.00	0.03	214	0.21	5,175,636	1,109,471	1109	55.47
20	83.83	179	2.07	5.00	0.04	228	0.23	5,447,358	1,239,951	1240	62.00
15	90.53	193	2.23	5.00	0.04	249	0.25	5,882,451	1,463,888	1464	73.19
10	99.53	212	2.45	5.00	0.04	277	0.28	6,467,215	1,793,963	1794	89.70

(continued)

Table 1.2 (continued)

Low curve Frequency (%)	At juntai (MCM/mth)	At dam (Mld)	At dam site m³/s	At dam site %	Low curve Q (m³/s)/km²	Concentration (mg/l)	Concentration (kg/m³)	Flow m³/month	Mass loading kg/month	Mass loading tonn/month	adj by fraction
5	114.75	244	2.83	3.00	0.05	326	0.33	7,456,082	2,428,076	2428	72.84
2	140.80	300	3.47	1.00	0.06	408	0.41	9,149,125	3,735,324	3735	37.35
1	169.41	361	4.18	1.00	0.07	499	0.50	11,007,841	5,492,658	5493	54.93
0.010	203.83	434	5.03	1.00	0.09	608	0.61	13,244,216	8,053,973	8054	80.54
				96.00						Total	806.43
											164

Note

Monthly flow duration curve is used

Total = 806.43 tonne/month * 12 month/59 km² = 164 $\underline{\text{tonnes/km}^2\text{/year}}$

Transposition of flow @ Juntai to dam site using areal ratio

Table 1.3 Sediment flow duration rating method: using high curve (Type II) and monthly flow duration curve

High curve Frequency	At juntai (MCM/mth)	At dam (Mld)	At dam site m³/s	At dam site %	High curve Q (m³/s)/km²	Concentration (mg/l)	Concentration (kg/m³)	Flow (m³/month)	Mass loading kg/month	Mass loading tonn/month	Mass loading adj by fraction
100	12.05	26	0.30	1.00	0.01	96	0.10	783,167	75,391.51	75	0.75
99	21.16	45	0.52	1.00	0.01	198	0.20	1,374,892	271,637	272	2.72
98	23.81	51	0.59	1.00	0.01	227	0.23	1,547,189	351,316	351	3.51
95	25.74	55	0.63	3.00	0.01	249	0.25	1,672,496	415,649	416	12.47
90	30.19	64	0.74	5.00	0.01	298	0.30	1,961,397	584,459	584	29.22
85	35.15	75	0.87	5.00	0.01	353	0.35	2,284,039	806,764	807	40.34
80	38.38	82	0.95	5.00	0.02	389	0.39	2,493,950	970,534	971	48.53
75	40.44	86	1.00	5.00	0.02	412	0.41	2,627,959	1,082,976	1083	54.15
70	43.80	93	1.08	5.00	0.02	449	0.45	2,846,347	1,279,394	1279	63.97
65	46.31	99	1.14	5.00	0.02	477	0.48	3,009,100	1,436,394	1436	71.82
60	49.51	105	1.22	5.00	0.02	559	0.56	3,216,878	1,798,305	1798	89.92
55	52.76	112	1.30	5.00	0.02	586	0.59	3,428,529	2,009,443	2009	100.47
50	55.73	119	1.37	5.00	0.02	610	0.61	3,621,093	2,210,147	2210	110.51
45	59.18	126	1.46	5.00	0.02	638	0.64	3,845,095	2,453,782	2454	122.69
40	61.95	132	1.53	5.00	0.03	660	0.66	4,025,476	2,657,807	2658	132.89
35	67.76	144	1.67	5.00	0.03	706	0.71	4,403,136	3,107,228	3107	155.36
30	72.06	154	1.78	5.00	0.03	739	0.74	4,682,157	3,458,293	3458	172.91
25	79.65	170	1.96	5.00	0.03	796	0.80	5,175,636	4,117,925	4118	205.90
20	83.83	179	2.07	5.00	0.04	826	0.83	5,447,358	4,501,880	4502	225.09
15	90.53	193	2.23	5.00	0.04	875	0.87	5,882,451	5,146,776	5147	257.34
10	99.53	212	2.45	5.00	0.04	939	0.94	6,467,215	6,070,752	6071	303.54

(continued)

Table 1.3 (continued)

High curve Frequency	At juntai (MCM/mth)	At dam (Mld)	At dam site m³/s	At dam site %	High curve Q (m³/s)/km²	Concentration (mg/l)	Concentration (kg/m³)	Flow (m³/month)	Mass loading kg/month	tonn/month	adj by fraction
5	114.75	244	2.83	3.00	0.05	1043	1.04	7,456,082	7,778,555	7779	233.36
2	140.80	300	3.47	1.00	0.06	1214	1.21	9,149,125	11,110,314	11,110	111.10
1	169.41	361	4.18	1.00	0.07	1393	1.39	11,007,841	15,334,311	15,334	153.34
0.010	203.83	434	5.03	1.00	0.09	1598	1.60	13,244,216	21,164,351	21,164	211.64
				96.00						Total	2912.79
											592

Note

Monthly flow duration curve is used

Total = 2912.79 tonne/month * 12 month/59 km^2 = 592 tonnes/km^2/year

Transposition of flow @ Juntai to dam site using areal ratio

Table 1.4 Sediment flow duration rating method: using low curve (Type 1) and daily flow duration curve

Low curve Frequency	At juntai (MCM/mth)	At dam (Mld)	At dam site m³/s	At dam site % interval	Low curve Q (m³/s)/km²	Concentration (mg/l)	Concentration (kg/m³)	Flow (m³/day)	Mass loading kg/day	Mass loading tonne/day	adj by fraction
0–0.1	259.01	552	6.39	0.10	0.11	763	0.76	551,810	420,919.46	421	0.42
0.1–0.5	228.55	487	5.64	1.40	0.10	687	0.69	486,911	334,270	334	4.68
0.5–1.0	192.55	410	4.75	3.51	0.08	572	0.57	410,222	234,798	235	8.24
1.0–5.0	142.59	304	3.52	6.99	0.06	414	0.41	303,779	125,746	126	8.79
5.0–12.0	99.00	211	2.44	8.00	0.04	276	0.28	210,922	58,157	58	4.65
12.0–20.0	57.08	122	1.41	14.01	0.02	143	0.14	121,601	17,363	17	2.43
20.0–34.0	55.84	119	1.38	21.00	0.02	139	0.14	118,963	16,519	17	3.47
34.0–55.0	44.61	95	1.10	18.00	0.02	103	0.10	95,047	9815	10	1.77
55.0–75.0	34.97	74	0.86	6.20	0.01	73	0.07	74,499	5414	5	0.34
75.0–80.0	29.36	63	0.72	5.90	0.01	55	0.05	62,541	3432	3	0.20
80.0–85.0	26.67	57	0.66	4.70	0.01	46	0.05	56,815	2634	3	0.12
85.0–90.0	24.19	52	0.60	4.50	0.01	39	0.04	51,537	1984	2	0.09
90.0–95.0	21.21	45	0.52	3.30	0.01	29	0.03	45,193	1313	1	0.04
95.0–98.0	17.05	36	0.42	2.00	0.01	16	0.02	36,323	576	1	0.01
98.0–99.0	16.26	35	0.40	0.26	0.01	13	0.01	34,639	462	0	0.00
99.0–99.9	9.07	19	0.22	0.12	0.00	−9	−0.01	19,312	−183	0	0.00
99.9–100.0	4.51	10	0.11	0.01	0.00	−24	−0.02	9600	−230	0	0.00
				99.90						Total	34.84
											216

Note

Daily flow duration curve is used

Total = 34.84 tonne/day * 365 days/59 km² = 216 tonnes/km²/year

Transposition of flow @ Juntai to dam site using areal ratio

Table 1.5 Sediment flow duration rating method: using high curve (Type II) and daily flow duration curve

High curve Frequency	At juntai (MCM/mth)	At dam (Mld)	At dam site m³/s	% interval	High curve Q (m³/s)/km²	Concentration (mg/l)	Concentration (kg/m³)	Flow m³/day	Mass loading kg/day	tonne/day	adj by fraction
0–0.1	259.01	552	6.39	0.10	0.11	1909	1.91	551,810	1,053,424.40	1053	1.05
0.1–0.5	228.55	487	5.64	1.40	0.10	1740	1.74	486,911	847,095	847	11.86
0.5–1.0	192.55	410	4.75	3.51	0.08	1532	1.53	410,222	628,434	628	22.06
1.0–5.0	142.59	304	3.52	6.99	0.06	1226	1.23	303,779	372,366	372	26.03
5.0–12.0	99.00	211	2.44	8.00	0.04	935	0.94	210,922	197,216	197	15.78
12.0–20.0	57.08	122	1.41	14.01	0.02	621	0.62	121,601	75,551	76	10.58
20.0–34.0	55.84	119	1.38	21.00	0.02	611	0.61	118,963	72,718	73	15.27
34.0–55.0	44.61	95	1.10	18.00	0.02	458	0.46	95,047	43,578	44	7.84
55.0–75.0	34.97	74	0.86	6.20	0.01	351	0.35	74,499	26,164	26	1.62
75.0–80.0	29.36	63	0.72	5.90	0.01	289	0.29	62,541	18,059	18	1.07
80.0–85.0	26.67	57	0.66	4.70	0.01	259	0.26	56,815	14,707	15	0.69
85.0–90.0	24.19	52	0.60	4.50	0.01	231	0.23	51,537	11,920	12	0.54
90.0–95.0	21.21	45	0.52	3.30	0.01	198	0.20	45,193	8956	9	0.30
95.0–98.0	17.05	36	0.42	2.00	0.01	152	0.15	36,323	5516	6	0.11
98.0–99.0	16.26	35	0.40	0.26	0.01	143	0.14	34,639	4955	5	0.01
99.0–99.9	9.07	19	0.22	0.12	0.00	63	0.06	19,312	1217	1	0.00
99.9–100.0	4.51	10	0.11	0.01	0.00	12	0.01	9600	118	0	0.00
				99.90						Total	113.76
											704

Note
Daily flow duration curve is used
Total = 113.76 tonne/day * 365 days/59 km² = 704 tonnes/km²/year
Transposition of flow @ Juntai to dam site using areal ratio

1.22 Results and Discussions

Two flow series (see Table 1.1) and both high and low sediment rating curves are used for estimating suspended sediment loads into the proposed Teriang reservoir (CA = 59 km^2). The monthly time step flows represent two different lengths of records, i.e. 1948 to 1975 (PWRS 1992) and 1948 to 2001 (augmented). Virtually no difference between the sedimentation loading for two different series of monthly inflow records used for the calculation. Table below shows the estimated suspended sediment loading.

Estimated Teriang Dam suspended sediment load

Flow records	No. of year	Total suspended sedimentation load tonne/(26 and 54 years respectively)		
1948–1975	26	CA = 59 km^2	High curve	Low curve
			932,268 (608 ton/km^2/year)	256,045 (167 ton/km^2/year)
1948–2001	54		1,955,741 (614 ton/km^2/year)	522,787 (164 ton/km^2/year)

The unit suspended sediment curves developed are used to derive annual sediment loading to the proposed Teriang reservoir. The low (Type II) and high (Type I) curves yield 167 and 608 tonne/km^2/year suspended sediment respectively. Assuming bed load equal to 20% of the suspended sediment (a reasonable assumption in line with past practices and confirmed in Lai [13] and adjustment made for different time interval, the total sediment load for both low and high curves are 261 and 948 tonne/km^2/year.

As mentioned earlier, the high curve (or Type I) represents groups of highly urbanized and significantly disturbed catchment sediment loading. This is certainly not the case for Sg. Teriang at the proposed dam site. At present it is still an undisturbed forested upland catchment. Therefore, it would be overwhelmingly as well as unduly conservative to adopt the "high curve" sediment flow scenario for Teriang dam basin in this Study. A middle approach would therefore perhaps sufficed and the adopted total sediment load of 250–300 tonne/km^2/year is therefore deemed reasonable. However by doing so, it is explicitly assumed that the catchment land use will remain the same as the sediment loads are estimated.

Alternative method is also used to check on the accuracy of sediment estimate using long-term flow method. This method utilizes both suspended sediment rating curve (Type I and II) and flow duration curves of monthly and daily time step of Sg. Teriang @ Juntai streamflow station (CA = 904 km^2). The results are comparable to the estimates using long-term flow records.

Annual average suspended sediment inflows

Time step	Low curve (tonne/km²/year)	High curve (tonne/km²/year)
Daily time step	216	704
Monthly time step	164	592
Time interval (daily/monthly) ratio	1.32	1.19
Adopted SS load daily	216	704
Adjustment for (hourly/daily) interval ratio × 1.20	259	845
Adjustment for bedload 20% of SS load	311	1014
Total sediment load	311 (261)	1014 (948)

(): calculated previously using long-term flow records

The difference between daily and monthly time interval is also explained. As expected using a smaller time step of daily average flow, the suspended load estimates are higher than that of using a monthly time step. The differences are 32 and 19% for low (Type I) and high (Type II) curves respectively.

1.23   Summaries and Conclusions

The review presented in this appendix by far is not exhaustive but it serves the purpose of estimating the reservoir sediment inflows. The limitations and constraints on predicting accurate sediment loading should be acknowledged a priori.

Estimated total sediment loads for proposed Teriang reservoir (CA = 59 km²) are 261 and 948 tonne/km²/year under low (Type II) and high (Type I) curves scenarios. The rationale of adopting the high curve (Type I) scenario does not seem to be consistent with the current land use patterns in Teriang reservoir catchment. On the other hand, if Type I low curve scenario is adopted, the estimated load is the same as the lowest limit of the sediment load advocated by SMHB in most of its dam/reservoir project.

Alternately Sediment flow duration rating approach is also used to calculate suspended sediment inflows into the Teriang reservoir. This approach uses both daily and monthly flow duration curves (see Fig. 1.3) and Type I and II sediment rating curves (see Figs. 1.1 and 1.2). The estimated total sediment loads are 311 and 1014 tonnes/km²/year for Type I and II curves respectively. These compare well with the earlier methodology, i.e. 261 and 948 tonnes/km²/year albeit some minor difference amounting to 7–16% considering the uncertainties in the sediment prediction. Overall, the sediment load computation for Type I curve agrees well with the SMHB convention of allocating some 260–300 tonnes/km²/year sediment inflows in its dam/reservoir design projects in Malaysia.

Therefore for 100-year deposition the sediment accumulation is 1.41 MCM if a sediment bulk density of 1.30 tonnes/m³ is assumed. It is also explicitly assumed

Unit = m3/s
CA= 904 km2
Flow record period = 1948 to 1975 with 1965-1969 discard due to error in gauging

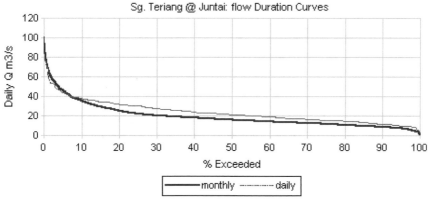

Minor difference between daily and monthly time step flow duration curves. Only difference is visible for 10% to 80% exceedance.

Fig. 1.3 Sg. Teriang @ Juntai Streamflow Station: monthly and daily flow duration curve (minor difference between daily and monthly time step flow duration curves. Only difference is visible for 10–80% exceedance)

that the dam/reservoir catchment management and good practices by prohibiting significant alteration of land use requirement should be recommended.

The limitations and insights of total sediment loadings estimates for proposed Teriang dam are discussed in order of importance as below:

1. The accuracy of total sediment load estimates into the proposed Teriang reservoir (CA = 59 km^2) based on the flow time interval used in the computation remains uncertain. It was based on a monthly time step in this Study. A coefficient of 1.20 that has been adopted in this Study was based on limited sampling carried out by SMEC [18] and SMEC/SMHB [17]. Balamurugam (1995) in his earlier paper in 1989 cited some underestimations with errors between 8 and 40% in sediment estimates if using mean daily discharges. To support this assertion, also cited in the article, Loughran (1976) quoted some 28–47% lower than that using hourly discharges. In this Study, the differences between monthly and daily time step in flow discharge are 1.32 and 1.19 respectively.

2. Assumption made on the bed load estimate as a fraction of observed suspended load (from 10 to 20% have been adopted in Malaysia) may not be adequate though with limited validations from field observation. Without further evidences on the concurrently observed ratio of bed load to suspended load, additional investigations are therefore required in this aspect.

3. The traditional sources of sediment records are provided mostly by JPS and to a certain extent, defunct Hydrology Unit of TNB. The sediment gauging programs

are hampered by logistic as well as financial constraints as the gauging program is extremely expensive to carry out vis-à-vis other hydrometric sampling program. Most of the time, the information obtained due to its scarcity and therefore fails to provide a meaningful and reasonable estimate of forest sedimentation rate

4. In the absence of elaborate and detailed information on sediment loadings, SMHB in the past in its assessment of water resources in both states of Pahang and Johor [20, 21] recommended a total sediment load of 260–300 tonnes/km^2/ year as preliminary estimates for respective reservoir scheme. It has to bear in mind that these recommended rates are deduced from experiences especially in the past studies. This is by far could be construed as the "average value" rate. On the other hand, this Study provides some detail calculations by taking into account of the temporal variations of both high and low regime in the flow discharges. The estimates reported in this Study nevertheless are consistent with the earlier provisional recommendation in the Pahang and Johor water resources studies [20, 21], which SMHB adopted for most of their dam/reservoir design projects.

5. The sediment gauging program in Malaysia is weak compared to other hydrometric program and lack of long-term records, if any mostly short term and with many missing gaps in between periods of records. Therefore efforts should be made to strengthening the gauging operation as it is not only an important aspect in reservoir/dam design but also on the sediment management in streams and rivers.

6. The computation of reservoir sediment loading using agriculture based hill slope type models, such as USLE and its variants (RUSLE, MUSLE, DUSLE [Europena version of USLE]) should be deemed with caution as these models are empirically based and their applications are theoretically conditioned to the impact of land development in agricultural hill slope areas. However, attempts are being made to establish the six factors (R, K, L, S, C, and P) of USLE for steep hillside slope to assess the erosion and land slide risks in highland Malaysia [23]. The original equation and intent of Wischmeir and Smith (1965) apparently is not suitable for estimating soil loss from drainage basins because it does not include crucial and appropriate factors, i.e. sediment delivery ratio, gully and streambank erosion, etc. As mentioned this simple multiplicative formula is developed based on extensive data collected in agricultural field plots. So its extrapolation to a different land use should also be deemed with caution. Also it is mostly based on long-term average results and it is therefore not suitable for predicting the soil movement due to single or individual storm events. Other recently developed genre of mechanistic and process-based hill slope erosion models, i.e. KINEROS, CALSITE, WEPP, SHE-TRANS etc. in addition to other "traditional USLE" type of CREAM, AGNPS, ANSWERS, and EPIC which maybe more suitable and appropriate for assessing event-based watershed soil erosions.

7. Inherent uncertainties and weaknesses in the sediment rating curves should be taken into consideration in the prediction of suspended load concentration using flow discharge as an independent variable. The rating curve should not be used outside the ranges of streamflow date from which it was derived. It is recommended by ICGB-ICOLD [4] that a continuous sampling or gauging program of at least five (5) years is needed to adequately cover the full range in flow discharges and to avoid significant extrapolation of the sediment rating curve. However a shorter period may be also possible if the range in both flow and sediment discharges is adequately covered for the purpose of including a single large event that may carry the equivalent of several normal years of sediment yield. This is explained earlier of applying a coefficient for sediment load calculation based on daily and shorted time interval.

8. It is believed that most of the larger size sediment will most likely be settled at the headwater upon entering the quiescent reservoir water body. It would be most unlikely that the sediment will completely fill up the assigned or allocated dead storage without inducing the sediment under current.

9. The sediment yields in the upper catchment basin, where most of the reservoir schemes located are estimated with a strict assumption that the dam/reservoir land uses remain the same as during their design stage. The impact and disturbance of the tropical forest with land use changes especially associated with human activities are well documented in researches carried out throughout the tropic countries. As such, the dam/reservoir catchment areas should therefore be kept the same as before via both structural and non-structural approaches. Sometimes, reforestation or afforestation help to restore the prevailing condition as before. There is also evidence of agro-forestry system (conversion of forest to oil palm crops) can more or less replicate the hydrological role of the original forest giving time for establishment (as quoted in [6] on FRIM studies on conversion of forested to oil palm land uses).

References

1. American Society of Civil Engineers. (1975). *Sedimentation engineering*. Manuals and reports on engineering practice No. 54. NY. NY.
2. Balamurugan, G. (1991). Some characteristics of sediment transport in the Klang River Basin. *Journal of the Institutions of Engineers Malaysia, 48,* 31–52.
3. Binnie dan Rakan. (1980). *Study on siltation at Pedas impounding dam*. Final Report, JBANS, Kerajaan Negeri Sembilan.
4. CIGB-ICOLD. (1989). *Sedimentation control of reservoirs: Guidelines*. Bulletin (Vol. 67), Paris, France.
5. Chan, N. W. (1998). Development of hill land and its effects on hydrology and water resources in Penang. In *Humid tropic*, November 24–26, 1998. Ipoh, Malaysia.
6. Douglas, I. (1999). Hydrological investigations of forest disturbance and land cover impacts in Southeast Asia: A review. *Philosophical Transactions of the Royal Society of London B: Biological Sciences, 354,* 1725–1738.

7. Douglas, I., Bidin, K., Balamurugan, G., Chappel, N. A., Walsh, R. P. D., Greer, T., et al. (1999). The role of extreme event in the impacts of selective tropical forestry erosion during harvesting and recovery phases at Danum Valley, Sabah. *Philosophical Transactions of the Royal Society of London B: Biological Sciences, 354,* 1749–1761.

8. Ekran Bhd. (1995). *Detailed environmental impact assessment: privatization of Bakun Hydroelectric project. Appendix 3A, physical environment of the Bakun catchment.* Final report.

9. Electrowatt and SMHB. (1993). *Murum hydroelectric project: Phase I report.* Sarawak Electricity Supply Corporation (SESCO), Sarawak.

10. Jurutera Jasa and SMHB. (2001). The proposed Gerugu dam, Sarikei, Sarawak. Final design report. Jabatan Kerja Raya, Sarawak

11. K.K. Projek Konsultant/SMHB Sdn. Bhd. (1992). *Kota Kinabalu water supply extension scheme: Stage 2.* Final report. JBA, Negeri Sabah.

12. Lai, F. S., & Akkharath, L. (2002). Suspended sediment yield changes resulting from forest harvesting in the Sg. Weng Experimental Watershed, Kedah, Peninsular Malaysia. In *International Conference on Urban Hydrology 2002*, October 14–16, 2002, Kuala Lumpur, Malaysia.

13. Lai, F. S. (1993). Sediment yield from logged, steep upland catchment in Peninsular Malaysia. In *Hydrology in warm humid regions*. IAHS Publication No. 216, pp 219–230.

14. McCuen, R. H. (1998). *Hydrologic analysis and design* (2nd ed.). New Jersey: Prentice Hall.

15. Mohd. Fuad, O., & Kamaruzaman, M. (1992). Mitigation of reservoir deposition through watershed management with special reference to hydropower scheme. In *Seminar on methods for preservation of useful reservoir storage on heavily sediment laden rivers*, October 12–16, 1992, Kuala Lumpur, Malaysia.

16. Mohd Amin, M. (1992). Partial restoration of reservoir capacity by periodic flushing and mechanical means. In *Seminar on methods for preservation of useful reservoir storage on heavily sediment laden rivers*, October 12–16, 1992, Kuala Lumpur, Malaysia.

17. SMEC and SMHB. (1988). Pergau hydroelectric project: feasibility study, vol 3: Hydrology. Final report, LLN and EPU, Government of Malaysia.

18. SMEC. (1976). *Terengganu River Basin study: Feasibility report on multi-purpose dam project*. Final Report (Vol. 3): Hydrology, EPU, Government of Malaysia.

19. SMHB/SMEC/SGV-KC. (1984). *Water resources development for East Negeri Sembilan, Melaka, Northeast Johor*. Final Report, EPU, Kerajaan Malaysia.

20. SMHB Sdn. Bhd. (1992). *Study on comprehensive water resources planning and development in the State of Pahang* (Vol 3: Chaps. 5 and 6). Final Report, EPU, Government of Malaysia.

21. SMHB Sdn Bhd. (1994). *Study of comprehensive water resources planning and development in the State of Johor*. Government of Johor: Final Report.

22. Sepakat Setia Perunding/Mott MacDonald. (1993). *Preliminary EIA, detailed engineering investigation, detailed design and supervision of construction of the Sg. Kelinchi dam and transfer tunnel*. Conceptual design Report, JBA, Negeri Sembilan

23. Tew, K. H., & Faisal, A. (2003). Near real time soul erosion risk assessment and an early warning system. *Bulletin Ingenieur, 19,* 21–25.

Notes: Secondary Sources

24. Peh, C. H. (1980). Runoff and sediment transport by surface wash in three forested areas of Peninsula Malaysia, Malay. *Forester, 43,* 56–67. Sg. Tekam catchment (CA = 0.57 km^2) 270 tonnes/km^2/year; forested land use.

25. Shallow, papers. (1956). No full reference. Mostly in the Cameron highland, Sg. Telom (CA = 77 km^2); Sg. Kial (CA = 21 km^2) and Sg. Bertam (CA = 73 km^2), the rate quoted was 300 tonnes/km^2/year.

Chapter 2
Analyzing Extreme Events Using Standardized Precipitation Index During the 20th Century for Surat District, India

N.R. Patel, T.M.V. Suryanarayana and D.T. Shete

Abstract The study of floods and droughts requires the knowledge of wet and dry event sequences. They are the two important extreme conditions which directly or indirectly affects every field of environmental science. These extreme conditions are due to the change in one, of the many but most important parameter, rainfall. The standardized precipitation index is designed to quantify the rainfall for multiple time scales. These time scales reflect the impact of drought/floods on the availability of the different water resources. The modified classification by Agnew is referred for the classification of wet and dry events during the 20th Century for Surat district. The monthly rainfall data from 1901 to 2000 is utilized to determine the SPI values. SPI was calculated for 4, 6, 12, 24 and 48 months time scales. The area experienced more than 20% years of dry and wet events for the 20th Century. It is observed that the years 1942, 1945 and 1959 are identified as severe wet events for all the time scale. Year 1998 is identified as moderate wet event for all time scale. Years 1936 and 1987 are identified as severe dry events and year 1935 is identified as moderate dry event for all the time scales. No extreme wet event was observed. For extreme dry scenario all the years identified for different time scales are different.

Keywords SPI · Wet events · Dry events

N.R. Patel · T.M.V. Suryanarayana (✉) · D.T. Shete
Faculty of Technology and Engineering, Water Resources Engineering and Management
Institute, The Maharaja Sayajirao University of Baroda, Samiala 391410, India
e-mail: drsurya-wremi@msubaroda.ac.in

N.R. Patel
e-mail: neha311081@yahoo.com

D.T. Shete
e-mail: dtshete@yahoo.com

© Springer Nature Singapore Pte Ltd. 2018
M. Majumder (ed.), *Application of Geographical Information Systems and Soft Computation
Techniques in Water and Water Based Renewable Energy Problems*, Water Resources
Development and Management, https://doi.org/10.1007/978-981-10-6205-6_2

2.1 Introduction

Floods and droughts are the two important aspects of hydrological hazard. Floods usually result either from heavy precipitation (rain or snow) or from rapid snowmelt or glacier discharge. Droughts are caused by dry weather conditions in which evaporation exceeds the available surface water. They are frequently characterised by water shortages. Understanding the causes and forecasting of heavy or scant precipitation and high evaporative demand (and hence of floods and droughts) form an important objective of any climate research scenario. These two types of extreme hydrological events are important to understand the climate..

Floods depend on many things such as climate, nature of the collecting basin, nature of the streams, soil, vegetative cover, amount of snow melt and overall rainfall. Annually, the Indian land mass receives rainfall of 88–89 cm with very high variation from region to region. In the state of Rajasthan, the rainfall is almost nil whereas in the state of Meghalaya, an average rainfall of 1000 cm occurred every year. So this variation in the occurrence of rainfall makes the country prone towards the situations like floods and droughts (http://nidm.gov.in/Chap6.htm).

Drought is a normal part of virtually every climate on the planet, even rainy ones. It is the most complex of all natural hazards, and it affects more people than any other hazard. It can be as expensive as floods and hurricanes. The impacts of drought are greater than the impacts of any other natural hazard. Droughts are periods of time when natural or managed water systems do not provide enough water to meet established human and environmental uses because of natural shortfalls in precipitation or streamflow.

The understanding that a deficit or excess of rainfall has different impacts on the ground water, reservoir storage, soil moisture, snowpack, and streamflow led McKee et al. [2] to develop the Standardized Precipitation Index (SPI). The SPI was designed to quantify the rainfall deficit for multiple time scales. These time scales reflect the impact of drought/floods on the availability of the different water resources. Soil moisture conditions respond to rainfall anomalies on a relatively short scale, while groundwater, streamflow, and reservoir storage reflect the longer-term rainfall anomalies. For these reasons, McKee et al. [2] originally calculated the SPI for 3, 6, 12, 24, and 48-month time scales.

The study by McKee et al. [2] showed the relationship between drought duration, drought frequency, and drought time scale using the Standardized Precipitation Index (SPI) given bywhere,

$$SPI = \frac{X_i - \overline{X}}{\sigma} \tag{2.1}$$

σ standardized deviation for the station,
X_i precipitation for the ith observation,
$\overline{X}$ mean precipitation for the station

The index has the advantages of being easily calculated, having modest data requirements, and being independent of the magnitude of mean rainfall and hence comparable over a range of climatic zones. It does, however, assume the data are normally distributed, and this can introduce complications for shorter time periods. Because precipitation is not normally distributed for time scales shorter than 12 months, an adjustment is made which allows the SPI to become normally distributed. Thus, the mean SPI for a time scale and a location is zero and the standard deviation is one. This is an advantage because the SPI is normalized so that wetter and drier climates can be represented in the same way. In addition, wet periods can also be monitored using the SPI.

A new classification presented in Table 2.1, for drought intensity has been proposed by Agnew [1] based on the Standardized Precipitation Index (SPI). This uses probability classes rather than magnitudes of the SPI for classification and is therefore suggested as a more rational approach. The effect is most noticeable at the demarcation of mild and moderate droughts. But purely statistical definitions of meteorological drought should be treated with caution. Perhaps of equal significance is the omission within the SPI of any assessment of persistence. It is rare that drought in any one year causes major hardship. It is the sequence of low rainfalls

Table 2.1 Classification of SPI values for wet and dry periods

SPI	Probability of occurrence	McKee et al. [2] Classification	Agnew [1] Classification
Less than −2.00	0.023	Extreme dry (drought)	
Less than −1.65	0.050		Extreme dry (drought)
Less than −1.50	0.067	Severe dry (drought)	
Less than −1.28	0.100		Severe dry (drought)
Less than −1.00	0.159	Moderate dry (drought)	
Less than −0.84	0.201		Moderate dry (drought)
Less than −0.50	0.309		No dry (drought)
Less than 0.00	0.500	Mild dry (drought)	
Less than 0.50	0.692		No wet
Less than 0.84	0.967		Moderate wet
Less than 1.00	0.159	Moderate wet	
Less than 1.28	0.841		Severe wet
Less than 1.50	0.933	Severe wet	
Less than 1.65	0.951		Extreme wet
Less than 2.00	0.977	Extreme wet	

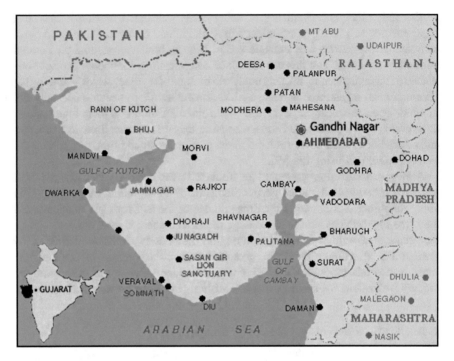

Fig. 2.1 Location of study area

that creates difficulties. For example, in England the drought of 1976 was really caused by the low rainfalls in the preceding year, while the drought of 1992 was the result of the low rainfalls from 1988. Thus the author concluded that the SPI needs to be developed from merely classifying intensities to include drought sequences, and the selection of appropriate averaging periods needs more attention.

2.2 Study Area

The present study analyses the occurrence of extreme events in the area considered for the study, i.e., Surat (Fig. 2.1) during the 20th century using the SPI for analyzing drought under various time scales and to determine the effectiveness of length of records for the same.

2.3 Methodology

Based on the methodology presented by Patel and Shete [3] best distribution is fitted to the precipitation dataset. The best distribution is selected using AIC and BIC criteria. The cumulative probability for the each rainfall event is then determined. The cumulative probabilities obtained are then transformed to the standard normal variate for determining SPI by Eq. 2.1. Both the rainfall and the SPI calculated are then plotted to study the behavior of SPI with respect to the rainfall values.

The SPI is calculated by taking the difference of the precipitation from the mean for a particular time scale, and then dividing by the standard deviation. The monthly rainfall data is then converted into 4, 6, 12, 24 and 48 months time scale. Thus SPI will be calculated for 4, 6, 12, 24 and 48. For determining the dataset using 24 and 48 months time scales the total of consecutive 12 months are considered. For example for SPI24 the total rainfall in 1902 is the sum of annual rainfall in 1901 and 1902. Similarly for 1903 the total rainfall is the sum of annual rainfall in 1902 and 1903. For SPI48 the total rainfall in 1904 is the sum of 1901, 1902, 1903 and 1904. Similarly for 1905 the total rainfall is the sum of 1902, 1903, 1904 and 1905. The time scales from 4 to 48 are considered for studying the short term as well as long term effect of rainfall. The wet and dry events are then classified based on Table 2.1 for observing the extreme events during 1901 to 2000 and the results are analyzed.

2.4 Results and Analysis

Based on AIC and BIC the best distribution fitted to the dataset is the inverse Gaussian and hence used for further analysis. The pdf for the distribution is given by

$$y = \sqrt{\frac{\lambda}{2\Pi x^3}} \exp\left\{ -\frac{\lambda}{2\mu^2 x}(x-\mu)^2 \right\}$$ (2.2)

where λ and μ are scale and shape parameters. The drought intensity is classified using the above methodology and the results are presented in Table 2.2.

From Tables 2.2 and 2.3 one can say that for SPI4 the total wet events were 29 and dry events were 22. For SPI6 the total wet events were 28 and dry events were 24. For SPI12 the total wet events were 29 and dry events were 23. For SPI24 the total wet events were 26 and dry events were 28. For SPI48 the total wet events were 26 and dry events were 28. One can say that as the time scale for the analysis increased the number of dry events exceeded the wet events.

It is observed that the years 1942, 1945 and 1959 are identified as severe wet events for all the time scale. Year 1998 is identified as moderate wet event for all time scale. Years 1936 and 1987 are identified as severe dry events and year 1935 is identified as moderate dry event for all the time scales. No extreme wet event was observed. For extreme dry scenario all the years identified for different time scales are different.

Figures 2.2, 2.3, 2.4, 2.5 and 2.6 represents the rainfall and SPI for time scales from 4 to 48. When the period of analysis is short, the variation between positive

Table 2.2 Wet events for Surat District based on SPI

SP I	Extreme	Severe	Moderate	Total events
4	No events	1914, 1942, 1945, 1946, 1947, 1954, 1959, 1964, 1970, 1976, 1983, 1994	1902, 1903, 1909, 1910, 1912, 1913, 1916, 1921, 1926, 1956, 1958, 1973, 1981, 1988, 1990, 1996, 1998	29
6	No events	1902, 1912, 1916, 1926, 1942, 1945, 1947, 1956, 1959, 1964, 1976, 1994	1903, 1921, 1931, 1939, 1944, 1946, 1954, 1958, 1973, 1975, 1981, 1983, 1988, 1990, 1996, 1998	28
12	No events	1942, 1945, 1946, 1959, 1970, 1976, 1983, 1994	1902, 1909, 1910, 1912, 1913, 1914, 1916, 1926, 1931, 1947, 1954, 1956, 1958, 1964, 1975, 1981, 1988, 1990, 1996, 1998,	28
24	No events	1913, 1914, 1917, 1943, 1945, 1946, 1947, 1959, 1976, 1984, 1994	1903, 1910, 1927, 1942, 1955, 1956, 1960, 1964, 1965, 1970, 1971, 1977, 1995, 1997, 1998	26
48	No events	1915, 1945, 1946, 1947, 1948, 1959, 1997	1914, 1916, 1917, 1942, 1944, 1949, 1956, 1957, 1958, 1960, 1961, 1976, 1978, 1984, 1991, 1994, 1996, 1998, 1999	26

Table 2.3 Dry events for Surat District based on SPI

SP I	Extreme	Severe	Moderate	Total events
4	1904, 1905, 1911, 1918, 1923, 1948, 1951, 1972, 1974	1901, 1915, 1920, 1936, 1982, 1987, 2002	1924, 1925, 1935, 1962, 1985, 1995	22
6	1911, 1918, 1925, 1974	1904, 1905, 1915, 1920, 1936, 1948, 1951, 1972, 1986, 1987	1901, 1923, 1930, 1935, 1952, 1957, 1978, 1980, 1982, 1999	24
12	1904, 1905, 1911, 1918, 1923, 1948, 1951, 1972, 1974	1901, 1915, 1920, 1936, 1982, 1987	1924, 1925, 1935, 1952, 1962, 1966, 1985, 1995	23
24	1905, 1924, 1952	1918, 1919, 1923, 1925, 1936, 1949, 1951, 1972, 1987	1906, 1911, 1935, 1937, 1953, 1962, 1963, 1966, 1969, 1973	28
48	1925, 1951	1906, 1907, 1920, 1921, 1923, 1924, 1926, 1937, 1938, 1952, 1953, 1974, 1975, 1987	1904, 1905, 1908, 1918, 1936, 1954, 1963, 1968, 1969, 1980, 1988, 1989	28

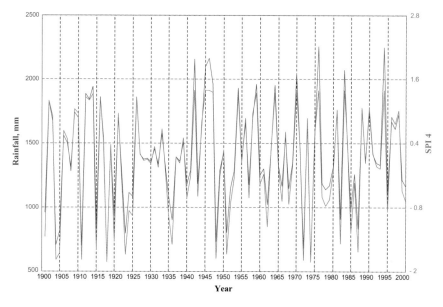

Fig. 2.2 Rainfall and SPI4 for Surat District

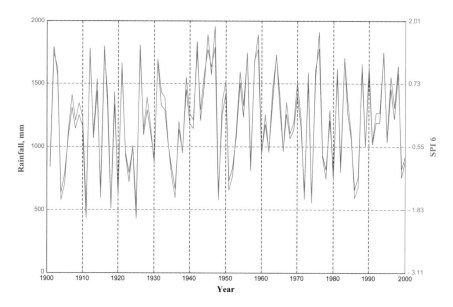

Fig. 2.3 Rainfall and SPI6 for Surat District

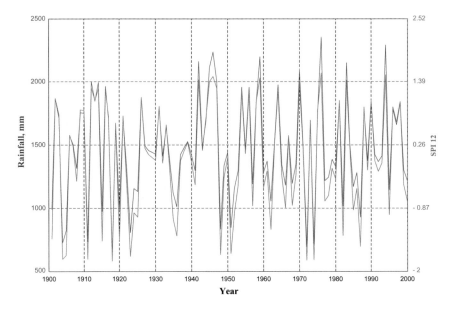

Fig. 2.4 Rainfall and SPI12 for Surat District

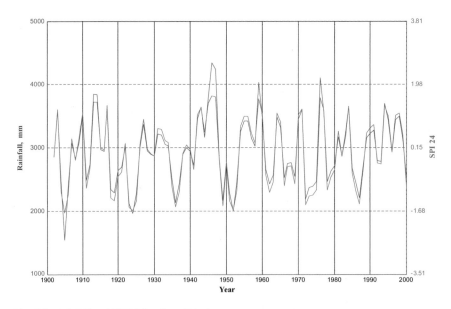

Fig. 2.5 Rainfall and SPI24 for Surat District

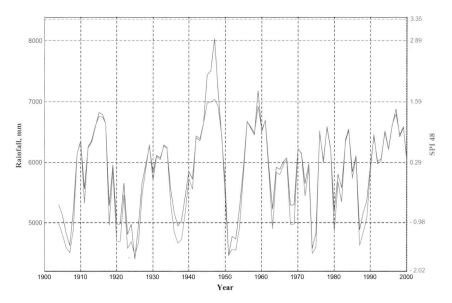

Fig. 2.6 Rainfall and SPI48 for Surat District

and negative values are seen more frequently and when the period of analysis increases, it is observed that the variation between positive and negative values are fewer. It is also observed that for 1000 mm rainfall the SPI values are 2.40, 0.36, 2.26, 1.83 and 1.30 for SPI4, SPI6, SPI12, SPI24 and SPI48 respectively. As SPI is directly proportional to the deviation of precipitation at any given unit of time from the mean precipitation, the SPI6 yields the better correlation with the rainfall.

2.5 Conclusions

The area under study experienced more than 20% years of dry and wet events for the 20th Century. As the time scale increases, the fluctuation from positive to negative or vice versa, decreases making the plot comparatively smooth. Also as the time scale for the analysis increased, the number of dry events exceeded the wet events. The superimposing of rainfall with SPI for various time scales leads to a conclusion that for the area taken under study, SPI6 yields better results, which means that the time scale of six months is to be preferred. Looking at this point, the dry events and wet events are 24 and 28 respectively. Thus the area has experienced comparatively more wet events.

Acknowledgements The rainfall data used for the analysis is obtained from the official website of India Water Portal, to whom the authors are very much thankful.

References

1. Agnew, C. T. (2000). *Using the SPI to identify drought.* DigitalCommons@University of Nebraska ñ Lincoln. http://digitalcommons.unl.edu/droughtnetnews/1.
2. McKee, T. B., Doesken, N. J., Kleist, J. (1993). The relationship of drought frequency and duration to time scales. Preprints. In: 8th conference on applied climatology, January 17–22, 1993, Anaheim, CA (pp. 179–184).
3. Patel, N. R., & Shete, D. T. (2008). Probability distribution analysis of consecutive days rainfall data for Sabarkantha District of North Gujarat Region, India. *Journal of Hydraulic Engineering (The Indian Society of Hydraulics), 14*(3), 43–55.

Chapter 3
An Approach to Develop an Alternative Water Quality Index Using FLDM

Ritabrata Roy

Abstract One of the major drawbacks of WQI is the manner, in which magnitude of the weights are assigned to the water quality parameters to their relative importance. Most of such methods are rather subjective and does not include the impact of the parameters to create hazards, cost to mitigate the hazard and utility to asses water quality. That is why; the present investigation proposes an objective method to determine the magnitude of weight, incorporating such factors and thus depicts the holistic quality of water. To achieve this objective FLDM was used to integrate the importance of parameters, on the basis of expert and literature survey, to represent actual situation of water quality more accurately. A case study of assessing the water quality of the sample water bodies in North East India with the proposed WQI is also incorporated to verify the applicability of the proposed index. The results show that the values of proposed WQI are close to that of NSF WQI and also in parity with available data.

Keywords Water quality index · WQI · Multi criteria decision making · MCDM · Analytical hierarchy process · AHP · Fuzzy · Surface water · Lentic · Tripura · India

3.1 Introduction

Water Quality Index (WQI) is a concise numerical representation of water quality which is convenient to interpret and used widely [31]. It is the weighted average of the concentration of the water quality parameters where weights are assigned to each of these parameters according to the goal and purpose of the index. However, selection of parameters to calculate WQI and to assign weights to them can be done in different ways [32], so WQI of a sample may vary, depending upon the methods

R. Roy (✉)
School of Hydro-Informatics Engineering, National Institute of Technology Agartala,
Barjala, Jirania, Tripura (W) 799055, India
e-mail: ritroy@gmail.com

© Springer Nature Singapore Pte Ltd. 2018
M. Majumder (ed.), *Application of Geographical Information Systems and Soft Computation Techniques in Water and Water Based Renewable Energy Problems*, Water Resources Development and Management, https://doi.org/10.1007/978-981-10-6205-6_3

adopted. The initial works to develop a WQI includes Horton's in 1965 [13] and Brown's in 1970 [4] (later known as NSF WQI), followed by Dunnette's Oregon Water Quality Index (OWQI) [9]. The NSF WQI is widely accepted and still being used extensively. Later many organizations throughout the globe developed several WQI for various purposes. The recently developed CCME WQI [5] and WAWQI [7] are more general in nature. In India, the Bhargava Water Quality Index [3] was developed to assess the quality of water in River Ganga.

In spite of being a convenient and useful technique, WQI is purpose specific, can cause loss and distortion of data [2, 32] and may fail to represent the general water quality [2] and aesthetic value [17] of water in an useful manner to substantiate complex technical decision making [23].

In recent years **Multi Criteria Decision Making (MCDM)** methods like **Analytical Hierarchical Process (AHP), Fuzzy Logic Decision Making (FLDM)** etc. are being widely used to support specific decision in complex situations, where multiple factors are involved [24].

FLDM is capable of converting vague qualitative statements (Fuzzy) into quantitative (Crisp) values [19]. So, the method is very suitable for indexing the water quality, which is a natural and qualitative entity. The suitability of FLDM to handle the uncertainty related to the expression of quality of water, which is a complex environmental attribute, is well tested [8]. The FLDM, therefore, can be used successfully to develop a suitable WQI [22].

3.1.1 Objective

The objective of the study is to propose a new index for determination of water quality of lentic water bodies, used mainly for agriculture, with FLDM incorporating the important criteria like Hazard Potential, Utility, Cost and Citation Frequency of the water quality parameters.

3.1.2 Study Area

Tripura, a state in North-East India, was selected for the case study. The state's main livelihood being cultivation, problems regarding the agricultural runoff are also common. Rivers and ponds are sources, as well as the sink, of the agricultural runoff [11]. Quality of agricultural runoff is considered to be important in maintaining the health and safety of the public.

The selected water quality parameters thirty ponds of Barjala area (Tripura West, India) were estimated. Ponds were selected by Belt Transect Survey [12]. Figure 3.1 shows the location of the sampling points. The WQI of the water of those ponds were worked out by the MCDM methods and the values were compared with those of a standard WQI for validation.

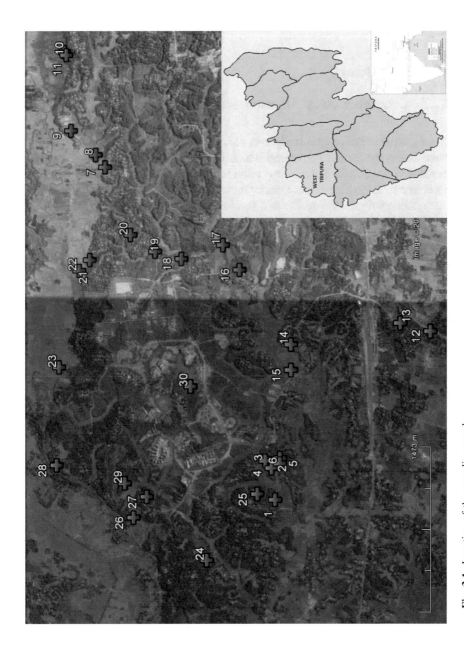

Fig. 3.1 Location of the sampling ponds

A case study of a location in the North East India where mainly canal (61% of total irrigated area) and tank (14% of total irrigated area) [6] fed irrigation are practiced.

3.2   Methodology

3.2.1   Development of Alternative Indices

The objective of the present study is to develop an alternative WQI, using MCDM tools. There are certain parameters which have influence on water quality. That means if Water Quality Parameter is referred to as P, and Quality of that parameter is referred to as Q, then Water Quality Index is a function of those parameters and their qualities. Again, the quality of a parameter depends upon its concentration (c).

WQI depends of Parameters and their Concentration

$$WQI = F(P, Q) \quad \text{and}$$
$$Q = f(c) \tag{3.1}$$

The parameters were identified by an extensive literature survey and discussions with the experts in related field.

3.2.1.1   Selection of MCDM Tool

In the present study the parameters can be represented by both qualitative and quantitative manner. Thus, **Fuzzy Logic Decision Making (FLDM)** method was adopted to find the weight of importance for each of the parameters which actually separates them according to their influence on overall water quality.

3.2.1.2   Method of Aggregation

WQI is a representation of overall water quality [33]. Thus, it should be an aggregation of various water quality parameters. As all the parameters do not influence the quality of water equally, there should be a suitable method of aggregation which can handle the importance of each parameter, as well as the importance of the concentration of each parameter.

Calculation of Weights of the Parameters

The main attributes of each of the water quality parameters are the weight of the parameter itself and the weight of the concentration of that parameter. As AHP and FLDM are used, suitable criteria and alternatives were to be worked out first for the calculation of the weights of the parameters.

Selection of Criteria

To find the weights of importance of the WQ parameters, some criteria have to be identified with respect to which the alternatives will be compared and the difference in importance can be determined. In this regard the following factors were considered for selection of Criteria:

(i) **Expert Survey**: A survey was carried out with experts of related fields where participants were asked to suggest those criteria which they consider important for determining water quality. The participants were also requested to provide their estimate about the most and least important criterion in this aspect. According to response received from the experts, scores were assigned to the criteria according to Eq. 3.2.

If A be number of experts who recommended a particular criterion, and a and b be the number of experts who referred to it as the most and the least important criterion respectively, then,
Importance by Expert Survey

$$S_E = \frac{A \times \left(\frac{a}{a'}\right)}{A_t} \tag{3.2}$$

where,

S_E Score of importance assigned to the criterion by Expert Survey
A   number of experts who recommended a particular criterion
a   number of experts, recommended that criterion most important
a'   number of experts, recommended that criterion least important
A_t Total number of Experts consulted

Where, S_E is the score assigned to the criterion and A_t is the total number of experts consulted regarding the present problem.

(ii) **Criteria Considered in Related Studies**: The literatures were also surveyed to find out the citation of the criteria in related studies. If the number of literatures which mentioned a criterion is c and the total number of literatures surveyed be C then the score, S_L, is calculated by Eq. 3.3.

Importance by Literature Survey

$$S_L = \left(\frac{c}{C}\right) \tag{3.3}$$

where,

S_L Score of importance assigned to the criterion by Literature Survey
c   Number of literatures which consider a particular criterion
C   Total number of literature surveyed

This score was also normalized and the parameters were ranked accordingly in a descending manner.

The significance score S was the normalized values of combination of these two scores.

Significance Score

$$S = \frac{S_E + S_L}{\sum (S_E + S_L)} \tag{3.4}$$

where,

S   Significance Score of a particular criterion
S_E Score of importance assigned to the criterion by Expert Survey
S_L Score of importance assigned to the criterion by Literature Survey

According to this final score, the most significant criteria were selected.

Selection of Alternatives

As the importance of the parameters were required to be estimated in the present study, all the parameters were considered as alternatives in FLDM method. The same process of scoring is followed for finding out their significance on the basis of expert survey and frequency of citation in literatures. The most significant parameters were selected to use in proposed WQI.

Calculation of Weights Using FLDM

FLDM method was utilized to determine the weights [34] of each of the parameters considered in the present study. The criteria were first compared with each other, based on their importance. In the present study, importance of each of the parameters is estimated in the following manner:

Criteria Ranking

The criteria were ranked according to the scores of their relative importance, calculated by Fuzzy pair wise comparison. The Fuzzy values were then converted to crisp values from Fuzzy membership graph. Thus a matrix of crisp values was generated from the matrix of Fuzzy values. The mean of the row of crisp values were taken as the score of the criteria.

Scoring of Alternatives

The scores of the alternatives were then calculated by pair wise comparison of the alternatives (i.e. the parameters) in respect to each of the criteria, using similar procedure.

Calculation of Weights of the Parameters

The weights of the different parameters were then calculated by multiplying the matrices of criteria scores and alternatives scores.

Calculation of Q Values

The Q value indices were developed from extensive expert survey on the basis of the effect of the concentration of a parameter on water quality (Brown et al. [4]. A greater Q value indicates better water quality, while a lesser Q value indicates worse.

3.2.1.3 Calculation of WQI

The FLDM WQI was then calculated by the weighted mean of the sub index values of the different parameters of a single sample. The sub index values were the products of the weights of respective parameters and their respective Q values [4].
Calculation of AHP WQI

$$WQI = \frac{\sum_{i=1}^{n} W_i Q_i}{\sum_{i=1}^{n} W_i} \qquad (3.5)$$

where,

Q_i Q value for ith water quality parameter
W_i weight associated with ith water quality parameter
n   number of water quality parameters

3.2.2 Validation of the Proposed WQI

For validation of the proposed WQI, two methods were applied—firstly, the WQI of the same samples were calculated using a standard WQI, and secondly, the results were compared with the WQI data available for the locality.

3.2.2.1 Comparison with Standard WQI

WQI of the same samples were calculated using an established and widely used method—the National Sanitation Foundation Water Quality Index (NSF WQI). The results from FLDM WQI were then compared with that from NSF WQI. The Mean Absolute Percentage Deviation (MAPD) [14] were calculated to depict the deviation of FLDM WQI from NSF WQI.

Calculation of MAPD of FLDM WQI from NSF WQI

$$MAPD = \frac{100}{n} \sum_{i=1}^{n} \frac{N_i - M_i}{N_i} \qquad (3.6)$$

where,

N_i NSF WQI of ith sample
M_i FLDM WQI of ith sample
n   Number of samples

3.2.2.2 Comparison with Available Data

The results from the proposed WQI were also compared with the available data of WQI of the water samples from the water bodies at the same locality.

3.2.3 Data Collection

The aim of the study was to find the potential of AHP in estimation of the quality of lentic water bodies by determining the WQI of the same. Ponds and rivers act as the sink of the runoff. That is why; samples were collected from the ponds of the study area, located in Barjala area, Jirania, West Tripura, India.

The concentrations of all the water quality parameters, except total hardness, were estimated in situ, using Sensor based Multi Water Parameter Analyzing devices. Biochemical Oxygen Demand (BOD) was measured in the laboratory by analytical method (EDTA Titrimetric Method [1]). Geographical locations of the sampling ponds were found by GPS device.

3.3 Results and Discussion

3.3.1 Calculation of Weights of the Parameters

The weights (i.e. importance) of the parameters were worked out from the relative ranking of the selected criteria and alternatives.

3.3.1.1 Selection of Criteria

Four criteria, viz. Hazard Potential (H), Utility (U), Cost (C) and Citation Frequency (F), were selected according to their significance score (Eq. 3.4).
Brief descriptions about the criteria are given below:

(i) **Hazard Potential (H)**: Hazard Potential is a relative estimation of the potential extent of hazard caused by each of the specific water quality parameters. The most hazardous (or least beneficial) parameter gets the lowest score.

The scores were assigned on the basis of occurrence in the literatures, in the following manner:
Calculation of the Importance of Hazard Potential

$$I_H = \frac{h}{C} \tag{3.7}$$

where,

I_H Importance of Hazard Potential
h   Number of times Hazard Potential given importance
C   Total number of times the Criteria were given importance

(ii) **Utility (U)**: Utility is a relative estimation of how useful each specific water quality parameter to represent the quality of the water. The most useful parameter gets the highest score.

The scores were assigned on the basis of occurrence in the literatures, in the following manner:
Calculation of the Importance of Hazard Potential

$$I_U = \frac{u}{C} \tag{3.8}$$

where,

I_U Importance of Utility
u   Number of times Utility given importance
C   Total number of times the Criteria were given importance

(iii) **Cost (C)**: It is a relative estimation of how expensive the mitigation of the negative impact imposed by each of the specific water quality parameters. The most expensive parameter gets the lowest score.

Calculation of the Importance of Hazard Potential

$$I_C = \frac{c}{C} \qquad (3.9)$$

where,

I_C Importance of Cost
c   Number of times Cost given importance
C   Total number of times the Criteria were given importance

(iv) **Citation Frequency (F)**: A metastatic analysis was carried out to find the parameters considered in the related studies which are already published in the reputed journals. A Citation Frequency (f_c) (Eq. 3.4), which represents the number of a parameter was considered in journal papers, was calculated for each of the parameters.

Calculation of Citation Frequency

$$f_c = \frac{p}{P} \qquad (3.10)$$

where,

f_c Citation Frequency
p   Number of times a single parameter occurs in a paper
P   Number of occurrence of all the parameters in papers surveyed

This Citation Frequency was considered as one of the representation of parameter importance in the present study.

3.3.1.2 Selection of Alternatives

In the present study water quality parameters were taken as the alternatives for the calculation of weights of those parameters. Eight parameters were selected for being most significant according to the literature survey.

(i)   Temperature
(ii)   pH
(iii) Turbidity
(iv)   Total Solids

   (v) Dissolved Oxygen
  (vi) Biochemical oxygen demand
 (vii) Total phosphate
(viii) Nitrates

3.3.1.3 Calculation of FLDM WQI

In the next step, the WQI of the sample water bodies were estimated, using FLDM WQI.

Ranking of Criteria

In FLDM the criteria were compared using Fuzzy values, which were later converted to crisp values (Chart 3.1; Table 3.1).

Scoring of Alternatives

The alternatives (water quality parameters) were compared on the basis of the importance against each of the criteria, using Fuzzy values. The Fuzzy values were later converted to crisp values to assign numerical scores to the parameters (Chart 3.2; Table 3.2).

Weights of the Parameters

Finally, the weights of the parameters were worked out by multiplication of the matrices of the scores of alternatives and criteria (Chart 3.3).

Chart 3.1 Ranking of Criteria in FLDM WQI

Table 3.1 Ranking of criteria according to fuzzy score

Criteria	Score	Rank
Hazard potential	0.7100	1
Utility	0.5000	2
Cost	0.3925	3
Citation frequency	0.2875	4

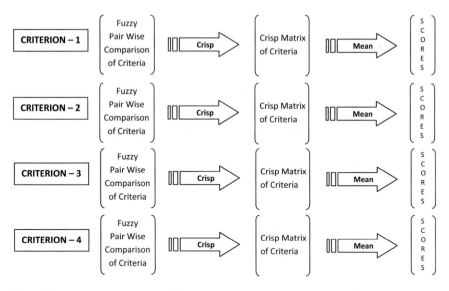

Chart 3.2 : Calculation of Scores of the Alternatives in FLDM WQI

Table 3.2 Scores of the alternatives (parameters) in FLDM WQI

Parameters	Scores			
	Hazard potential	Utility	Cost	Citation frequency
Dissolved oxygen	0.7463	0.7738	0.7600	0.6913
pH	0.5375	0.5175	0.1700	0.7188
BOD	0.5025	0.4825	0.6625	0.6313
Temperature	0.2688	0.5625	0.5900	0.5413
Total phosphate	0.5138	0.4475	0.4263	0.6838
Nitrates	0.5138	0.4475	0.4575	0.4750
Turbidity	0.5063	0.3913	0.3838	0.2225
Total solids	0.3313	0.3563	0.5825	0.3700

$$\begin{pmatrix} \text{Matrix of} \\ \text{Scores of the} \\ \text{Alternatives} \end{pmatrix} \times \begin{pmatrix} \text{Matrix of} \\ \text{Scores of} \\ \text{the Criteria} \end{pmatrix} = \begin{pmatrix} \text{W} \\ \text{E} \\ \text{I} \\ \text{G} \\ \text{H} \\ \text{T} \end{pmatrix}$$

Chart 3.3 Calculation of Weights of the Parameters in FLDM WQI

The corrected weights of FLDM WQI were calculated using Eq. (3.10) in the same manner as in AHP WQI.

These corrected weights of the parameters were used to calculate the FLDM WQI of the sample water bodies.

3.3.2 Comparison of FLDM WQI with NSF WQI

The results of the calculation of WQI, using the two different methods, are given below. The results were presented both in tabular form and in graph (Table 3.3).

The interpretation of the WQI values into human conceivable statement was done as per NSF WQI guideline (Table 3.4).

The values of WQI from all the three methods are presented graphically (Graph 3.1) to visualize the comparison between the two WQI.

It can be observed from the results that the WQI of the sample water bodies ranges from 47 to 73 with a mean of 64 and a standard deviation of 6.49 (Table 3.5). So, it can be said that the WQI of the sample have little variation (Table 3.6).

Table 3.3 Weights of the parameters in FLDM WQI

Parameters	Alternatives score				Criteria score	Parameter weights
	Hazard potential	Utility	Cost	Citation frequency		
Dissolved oxygen	0.7463	0.7738	0.7600	0.6913	0.7100	0.1860
pH	0.5375	0.5175	0.1700	0.7188	0.5000	0.1202
BOD	0.5025	0.4825	0.6625	0.6313	0.3925	0.1368
Temperature	0.2688	0.5625	0.5900	0.5413	0.2875	0.1130
Total phosphate	0.5138	0.4475	0.4263	0.6838		0.1253
Nitrates	0.5138	0.4475	0.4575	0.4750		0.1190
Turbidity	0.5063	0.3913	0.3838	0.2225		0.1013
Total solids	0.3313	0.3563	0.5825	0.3700		0.0984

Table 3.4 Corrected weights of the parameters in FLDM WQI

Parameters	Corrected weights
Dissolved oxygen	0.17
pH	0.11
BOD	0.11
Temperature	0.10
Total phosphate	0.10
Nitrates	0.10
Turbidity	0.08
Total solids	0.07

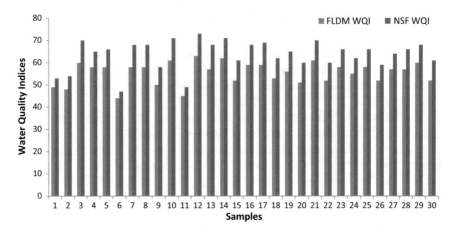

Graph 3.1 Comparison of Water Quality Indices by AHP, LDM and NSF Methods

According to NSF WQI, most of the sample ponds (80%) had a medium water quality, a few (13.33%) had good water quality and only a very few (6.67) had a bad water quality. However, according to FLDM WQI, a greater number of ponds (86.67%) had medium water quality and a few (13.33) had bad water quality and no pond had good water quality (Table 3.7).

The results show that FLDM WQI deviates 12.5% (Table 3.8) from the NSF WQI (Eq. 3.6) and FLDM WQI interpretation deviates from that of NSF WQI for six (20%) samples. Therefore, it can be said that values of FLDM WQI are quite close to that of NSF WQI.

However, the average water quality is interpreted as **Medium** by both of the methods. All these results indicate suitability of FLDM WQI, used in this study, in calculation of WQI.

3.3.3   Comparison of Results with Available Data

Available water quality data of other major water bodies of South Tripura district (Table 3.9) are also in parity with the WQI of the samples by all the methods used in this study.

This serves as further validation of the integrity of the MCDM WQI.

Table 3.5 Comparison of FLDM WQI with NSF WQI

Sample No.	FLDM method		NSF method	
	WQI	Interpretation	WQI	Interpretation
1	49	Bad	53	Medium
2	48	Bad	54	Medium
3	60	Medium	70	Medium
4	58	Medium	65	Medium
5	58	Medium	66	Medium
6	44	Bad	47	Bad
7	58	Medium	68	Medium
8	58	Medium	68	Medium
9	50	Medium	58	Medium
10	61	Medium	71	Good
11	45	Bad	49	Bad
12	63	Medium	73	Good
13	57	Medium	68	Medium
14	62	Medium	71	Good
15	52	Medium	61	Medium
16	59	Medium	68	Medium
17	59	Medium	69	Medium
18	53	Medium	62	Medium
19	56	Medium	65	Medium
20	51	Medium	60	Medium
21	61	Medium	70	Good
22	52	Medium	60	Medium
23	58	Medium	66	Medium
24	55	Medium	62	Medium
25	58	Medium	66	Medium
26	52	Medium	59	Medium
27	57	Medium	64	Medium
28	57	Medium	66	Medium
29	60	Medium	68	Medium
30	52	Medium	61	Medium
Mean	55	Medium	64	Medium
SD	4.97		6.49	

Table 3.6 NSF WQI scale for water quality for ranges of WQI values

Range	Quality
90–100	Excellent
70–90	Good
50–70	Medium
25–50	Bad
0–25	Very bad

Table 3.7 Water qualities of the sample water bodies

Water quality	Percentage of sample water bodies	
	NSF WQI	FLDM WQI
Good	13.33	0.00
Medium	80.00	86.67
Bad	6.67	13.33

Table 3.8 MAPD of FLDM WQI from NSF WQI

MAPD	12.50%

Table 3.9 Available of water quality data of water bodies in the same locality

SL	Water body	Location	NSF WQI (mean)	Interpretation
1	Amar Sagar	Udaipur, South Tripura	68	Medium
2	Dhani Sagar	Udaipur, South Tripura	60	Medium
3	Mahadeb Dighi	Udaipur, South Tripura	74	Good
4	Kalyan Sagar	Udaipur, South Tripura	76	Good
5	Hrishidas Colony Pond	Pratapgarh, South Tripura	52	Medium
6	Unnamed Dighi	Melarmath, Agartala, South Tripura	56	Medium
Average			64.33	Medium

Source Reports Published By ENVIS Centre, Tripura State Pollution Control Board, Tripura, India

3.4 Conclusion

In the present study, an alternative objective WQI was tried to develop using FLDM method, taking Hazard Potential, Utility, Cost and Citation Frequency as criteria and 8 selected water quality parameters as alternatives. The selected water quality parameters of 30 water bodies in West Tripura were worked out and their WQI were calculated using the proposed method. The WQI of the same water bodies were also calculated by NSF WQI, which is a widely accepted, well established method, for validation of the proposed method. Further, the results were compared with the available data of WQI of the various water bodies in the same locality.

The results of the study show that the average WQI value by FLDM Method and that by NSF Method differs a little. However, the interpretations of average WQI for both the methods are same (medium). Again, the average WQI of FLDM and NSF were very close to that of local water bodies as found in other studies. Also, the interpretation of the WQI of the local water bodies is same (medium) in both of the methods used in this study.

So, it can be concluded that the proposed method of calculation of WQI using FLDM is a valid method. The study suggests that the lentic water quality of the study area is medium in general.

Acknowledgements The authors thankfully acknowledge the active help of Mr. Santanu Bhowmik, Research Scholar, School of Hydro-Informatics Engineering, NIT Agartala during the collection of data for the present study.

References

1. American Public Health Association, American Water Works Association, Water Environment Federation. (1999). *Standard methods for the examination of water and wastewater* (20th ed.); 2340 C.
2. Ansari, K., Hemke, N. M. (2013). Water quality index for assessment of water samples of different zones in Chandrapur City. *International Journal of Engineering Research and Applications (IJERA)*, *3*(3), 236. ISSN: 2248-9622, May–Jun 2013.
3. Bhargava, D. S. (1983). Use of water quality index for river classification and zoning of Ganga River. Environmental Pollution (Series B)*, 6*, 51–67.
4. Brown Robert, M., McClelland Nina, I., Deininger Rolf, A., & Tozer Ronald, G. (1970). A water quality index—do we dare? *Water and Sewage Works*, 339–343, October 1970
5. Canadian Council of Ministers of the Environment. (2001). *Canadian environmental quality guidelines for the protection of aquatic life.* CCME Water Quality Index: Technical Report, 1.0.
6. Central Water Commission. (2005). *Water data complete book 2005* (pp. 67, Table: 2.12). http://www.cwc.nic.in/main/downloads/Water_Data_Complete_Book_2005.pdf.
7. Chauhan, A., & Singh, S. (2010). Evaluation of Ganga water for drinking purpose by water quality index at Rishikesh, Uttarakhand, India. *Report Opinion, 2*(9), 53–61.
8. Chen, H.W., Chang, N.B. (2001) Identification of river water quality using the fuzzy synthetic evaluation approach. *Journal of Environmental Management, 63*(3), 293–305.
9. Dunnette, D. A. (1979). A geographically variable water quality index used in Oregon. *Journal of the Water Pollution Control Federation, 51*(1), 53–61.
10. Forman, E. H., Saul, I. G. (2001). The analytical hierarchy process—an exposition. *Operations Research, 49*(4), 469–448.
11. Heal, K. V., Vinten, A. J. A., Gouriveau, F., Zhang, J., Windsor, M., D'Arcy, B., Frost, A., Gairns, L., Langan, S. J. (2006). The use of ponds to reduce pollution from potentially contaminated steading runoff. In *Agriculture and the Environment—Managing Diffuse Agricultural Pollution, Proceedings of the SAC and SEPA Biennial Conference*, April 5–6, 2006, Edinburgh (pp. 62–70).
12. Hill, D. A., Fasham, M., Tucker, G., Shewry, M., Shaw, P. (2005). *Handbook of biodiversity methods: survey, evaluation and monitoring* (pp. 219–222). Cambridge University Press.
13. Horton, R. K. (1965) An index number system for rating water quality. *Journal of the Water Pollution Control Federation, 37*(3), 300–305.
14. Hyndman, R. J., Athanasopoulos, G. (2013). *Forecasting: principles and practice.* OTexts.
15. Kasperczyk, N., Knickel, K. (2006). *The analytic hierarchy process (AHP).* http://www.ivm.vu.nl/en/Images/MCA3_tcm53–161529.pdf.
16. Macharis, C., Springael, J., De Brucker, K., Verbeke, A. (2004). Promethee and AHP: the design of operational synergies in multicriteria analysis. Strengthening Promethee with ideas of AHP. *European Journal of Operational Research, 153*, 307–317.

17. Mamun, A. A., Idris, A. (2008). Revised water quality indices for the protection of rivers in Malaysia. In *Twelfth International Water Technology Conference, IWTC12 2008*, Alexandria, Egypt (p. 1690)
18. McClelland, N. I. (1974). Water quality index application in the Kansas River Basin. EPA-907/9-74-001; US EPA Region VII, Kansas City, MO
19. McNeil, F. M., Thro, E. (1994). Fuzzy logic: A practical approach (p. 294). Boston: Academic Press.
20. Millet, I., Wedley, W. C. (2002). *Modelling risk and uncertainty with the analytic hierarchy process. Journal of Multi-Criteria Decision Analysis, 11*, 97–107.
21. Oram, B. (undated). *Monitoring the quality of Surfacewaters*. B.F. Environmental Consultants Inc. http://www.water-research.net/watrqualindex/.
22. Raman, B. V., Reinier, B., & Mohan, S. (2009). Fuzzy logic water quality index and importance of water quality parameters. *Air, Soil and Water Research, 2*, 51–59.
23. Ramanathan, R. (2001). A note on the use of the analytic hierarchy process for environmental impact assessment. *Journal of Environmental Management, 63*, 27–35.
24. Triantaphyllou, E. (2000). Multi-criteria decision making: a comparative study. Kluwer Academic Publishers (now Springer).
25. Tripura State Pollution Control Board. (Undated). *Assessment of pollution status of Dhani Sagar, Udaipur, South Tripura*. http://trpenvis.nic.in/test/doc_files/Dhani%20sagar.doc.
26. Tripura State Pollution Control Board. (Undated). Environment and ecological assessment of Kalyan Sagar, Udaipur. http://trpenvis.nic.in/test/doc_files/Kalyansagareport.pdf.
27. Tripura State Pollution Control Board. (Undated). *Pollution status assessment of Amarsagar, South Tripura*. http://trpenvis.nic.in/test/doc_files/Amarsagar.pdf.
28. Tripura State Pollution Control Board. (Undated). *Pollution status assessment of Mahadeb Dighi, South Tripura*. http://trpenvis.nic.in/test/doc_files/Mahadeb%20Dighi.pdf.
29. Tripura State Pollution Control Board. (Undated). *Pollution status assessment of Hrishidas Colony Pond, Pratapgarh; Undated. Pollution Status Assessment of Rana Deghi*. http://trpenvis.nic.in/test/doc_files/Rishidas%20Colonyreport.pdf.
30. Tripura State Pollution Control Board. (Undated). *Pollution status assessment of a Deghi situated on the North of Agiye Chala Sangha, Melarmath, Agartala*. http://trpenvis.nic.in/test/doc_files/Introduction.doc.
31. Tyagi, S., Sharma, B., Singh, P., Dobhal, R. (2013). Water quality assessment in terms of water quality index. *American Journal of Water Resources, 1*(3), 34–38.
32. Walsh, P., & Wheeler, W. (2012). *Working paper: Water quality index aggregation and cost benefit analysis*. NCEE Working Paper Series; National Center for Environmental Economics, U.S. Environmental Protection Agency.
33. Yogendra, K., & Puttaiah, E. T. (2007). Determination of water quality index and suitability of an urban waterbody in Shimoga Town, Karnataka. In *Proceedings of Taal 2007: The 12th World Lake Conference* (p. 342).
34. Zadeh, L. A. (1965). Fuzzy sets. *Information and Control, 8*, 338–353.

Part II
Water Based Renewable Energy Problems

Chapter 4
Development of Financial Liability Index for Hydropower Plant with MCDM and Neuro-genetic Models

Priyanka Majumder and Apu Kumar Saha

Abstract The population overgrowth along with technological advancement has increased the demand for Electricity all over the World. The cost of Electricity has been increased simultaneously. As the resources of conventional fuels are limited, alternate energy sources to replace fossil fuels are now preferred to supply the excess demand. Among all the sources of alternate energy, hydropower was found to be the most reliable but inexpensive source of alternative Energy. But locational implications and variation in kinetic Energy of water flow during the monsoon and non-monsoon seasons attracts sufficient amount of financial liability. Thus for any hydropower projects the financial liability are evaluated before approving the installation of the project. The conventional practices of liability analysis give equal importance to all the considered factors. But in reality not all the factors have the same importance on liability analysis of Hydropower projects. Thus some factors are overrated and some other under rated which resulted in erroneous decision making. The present investigation proposed a new method of liability analysis where all the major factors were given separate importance as decided from literature, Expert and local surveys. A financial liability index was also proposed to represent the financial liability of the project. The index was applied to hydro-power plants of different capacity and efficiency. The results are found to be coherent with the actual scenario. The index utilized various MCDM techniques followed by ANN architecture to create a flexible but cognitive instrument to analyse the financial liability of new hydro power project.

Keywords Hydropower plant · AHP · ANP · Fuzzy MCDM · FANP · ANN

P. Majumder (✉) · A.K. Saha
Department of Mathematics, National Institute of Technology Agartala, Agartala, India
e-mail: majumderpriyanka94@yahoo.com

A.K. Saha
e-mail: apusaha_nita@yahoo.co.in

© Springer Nature Singapore Pte Ltd. 2018
M. Majumder (ed.), *Application of Geographical Information Systems and Soft Computation Techniques in Water and Water Based Renewable Energy Problems*, Water Resources Development and Management, https://doi.org/10.1007/978-981-10-6205-6_4

4.1 Introduction

According to a World Bank report hydropower is treated as the most inexpensible and reliable form of alternative energy among all the available source of renewable energy (World Bank). Hydropower plants convert the potential energy of water column into the kinetic energy of the turbine which produce electricity by rotating the attached synchronators.

4.1.1 Hydro Power Plant (HPP)

Hydro power was classified by their size. Let x_m be areas away from the grid. A hydro power plant said to be large-hydro if $x_m \geq 100\,MW$ (megawatt), medium-hydro if $15\,MW \leq x_m < 100\,MW$, Small-hydro if $1\,MW \leq x_m < 15\,MW$, Mini-hydro if $1\,MW < x_m$ and $x_m > 100\,kW$ (kilowatt), Micro-hydro if $5\,kW < x_m \leq 100\,kW$ and Pico-hydro if $x_m \leq 5\,kW$.

Here hydroelectric power plants based on the height of water available in the reservoir, these are: Let y_m be height of water available in the reservoir. A hydro power plant said to be low head hydro if $y_m \leq 15\,m$ (meters), medium head hydroelectric power plants if $15\,m < y_m \leq 70\,m$, high head hydroelectric power plants if $70\,m < y_m \leq 250\,m$, Very high head plants if $y_m \geq 250\,m$.

Again hydro power was classified by their capacity. Let z_m be capacity of HPP. A hydro power plant said to be very low capacity hydroelectric plants if $z_m \leq 0.1\,MW$ (megawatt), medium capacity hydroelectric plants if $1.0\,MW \leq z_m < 10\,MW$, high capacity hydroelectric plants if $z_m \geq 10\,MW$.

4.1.2 Financial Factors of HPP

The importance of each cost factor is again different with respect to the other cost factor. Not all the factors are equally important. This importance can be estimated with the help of literature review and a survey among the experts of the industry. Overall cost factors can maintenance and operation cost are highly dependent on the location and capacity, capital cost dependent on the location, total production cost dependent on the location and capacity, project cost dependent on the location, energy equipment cost dependent on the capacity, plant cost dependent on the capacity, civil work cost dependent on the location, turbine cost dependent on the location and capacity, investment cost highly dependent on the location, initial investment cost highly dependent on the location, fuel cost dependent on the capacity and labour cost dependent on the location.

Table 4.1 depicts the cost factors and percentage of papers, expert and local survey which cite the said type of cost was depicted so that importance of the same over the others can be estimated in an objective manner. According to the analysis it was found that Maintenance and Operation Cost mostly impacts on the profitability of any hydropower plant whereas when Labour Cost is only considered it has minimum influence on the overall expenditure incurred in a HPP.

(a) Maintenance and operation cost (m) is

$$m = (\frac{u}{v} - \frac{a}{v(p \times o)}), \tag{4.1}$$

(b) Capital cost (c) is

$$c = \frac{d(1 - (1+i)^{-n})}{i}, \tag{4.2}$$

(c) Total production cost (T) is

$$T = p_v - r, \tag{4.3}$$

(d) Project cost (j) is

$$j = \frac{\sum_l k_l}{L}(1 - e), \tag{4.4}$$

(e) Energy equipment cost (E) is

$$= (\sum_{t=0}^{s} \frac{(I_t + m_t)}{(I_0 + R)^t})(\sum_{t=0}^{s} \frac{F_t}{(I_0 + R)^t})^{-1}, \tag{4.5}$$

Table 4.1 Table showing percentage of papers, expert and local survey

Cost factor	Papers (%)	Expert (%)	Local (%)
Maintenance and operation cost	86.36	80	88.33
Capital cost	63.64	64.12	59.45
Total production cost	54.54	57.18	60.27
Project cost	50	47.72	53
Energy equipment cost	45.45	48.33	50.91
Plant cost	40.91	45.64	42.84
Civil work cost	36.36	40	38.55
Turbine cost	31.82	34.91	37.34
Investment cost	30.28	33.36	28.78
Initial investment cost	27.27	25.45	23.23
Fuel cost	22.72	20.28	23
Labour cost	18.18	15.54	25

(f) Plant cost (y) is

$$y = f_1(\beta_1, \omega, \partial, \beta_2),$$ (4.6)

(g) Civil work cost (x) is

$$x = \sum_{hg} \sigma_{hg},$$ (4.7)

(h) Turbine cost (b) is

$$b = f_2(\alpha),$$ (4.8)

(i) Investment cost (β) is

$$\beta = f_2(\hbar, \varphi, \mathtext{ч}),$$ (4.9)

(j) Initial investment cost (δ) is

$$\delta = \frac{\mu - \tau}{\theta},$$ (4.10)

(k) Fuel cost (H) is

$$H = f_3(G),$$ (4.11)

(l) Labour cost (A) is

$$A = f_4(\sum ş, ӌ, Ӡ)$$ (4.12)

where u denotes unit production cost, v denotes production volume, a denotes annual debt service, p denotes plant design capacity, o denotes plant operating factor, d denotes annual debt service, i denotes annual interest rate for capital borrowing, n denotes number of year to repay the debt, p_v denotes total production value, r denotes net return, k denotes item cost per lane Km$_l$, e denotes engineering contingency %, L denotes cumulative principal cost items percentage, I_t denotes investments at time t, m_t denotes operation and maintenance costs at time t, F_t denotes energy generation at time t, R denotes the evaluation discount rate, β_1 denotes indirect labour cost, ω denotes indirect cost; ∂ denotes costs involved in manufacturing with the exception of the cost of raw materials, β_2 denotes direct labour, σ_{hg} denotes price of the number g subentry engineering of the number h branch engineering, α denotes Cost required for generate electricity, $\hbar$ denotes made through in-house capital, φ denotes credit from national financing agencies, ч denotes international financing agencies, μ denotes annual life cycle cost, τ denotes annual expenses, θ denotes capital recovery factor, G denotes all gas and electricity costs those are involved to run a project, ş denotes paid to employees, ӌ denotes cost of employee benefits and Ӡ denotes payroll taxes.

4.1.3 Justification and Objective

As discussed in the previous sections there are various types of HPP and each type has its own financial liability. But not all the financial factors are equally hazardous to the profitability of a plant. Various types of HPP will have different type of financial factors which can effect the plant income. The present investigate has three objectives. The first objective is to find the most important cost factor which can induce maximum effect on the plant income. The second objective is to find the least sensitive financial instrument HPP and finally the last objective of the study is to create a cognitive fool which can analyse the financial liability of any HPP based on the importance of each factor on the profitability of the proposed project.

The main objective of the study was three fold. First objective is to find the most important cost factor of HPP which can be introduced considerable liability upon the project. The second objective is to find the least important factor which can be neglected during financial evolution of a project. Lastly, the final objective is to develop a tool which can automatically predict the financial liability of any project once the magnitude of the cost factors are given. This tool will enable professional for a fast evaluation of existing as well as new project proposals. As the tool predicts the output based on the liability index developed w.t. help of the factors and the weights which separates the cost option into different level of importance. Thus making the result cognitive and objective without any aesthetical influence.

4.2 Multi Criteria Decision Making Methods

Multiple criteria decision making (MCDM) is making decisions in the presence of multiple, conflicting, criteria. MCDM problems are common in everyday life. In personal, one buys a house or a car may be characterized in terms of price, size, style, safety, comfort, etc. In business, MCDM problems are harder and of large scale. For example, many companies in Europe are conducting organizational self-assessment using hundreds of criteria and sub-criteria set in the EFQM (European Foundation for Quality Management) business excellence model. Purchasing departments of large companies often need to evaluate their suppliers using a range of criteria in different area, such as after sale service, quality management, financial stability, etc. The history of MCDM is of 30 years. The development of MCDM is related to development of computer technology. In one hand, the rapid development of computer technology in recent years has made it possible to conduct systematic analysis of complex MCDM problems. On the other hand, the widespread use of computers and information technology has generated a huge amount of information, which makes MCDM increasingly important and useful in supporting business decision making. Many methods are there to solve MCDM problems. It was reviewed by Hwang and Yoon (1981), though some of the methods were criticized as ad hoc and to certain degree unjustified on theoretical and/or empirical grounds. In early 1990s, new methods that could produce consistent and rational results, able to deal with uncertainties and provide transparency to the analysis processes were called to introduce. The great advantage of MAUT (multi attributes utilization techniques) is the uncertainty which it brings under consideration. But it is not a quality that is considered in many MCDM methods. It is comprehensive and can account for and incorporate the preferences of each consequence at every step of the method. This amount of accuracy is convenient; however besides this advantage it has many possible disadvantages. An incredible amount of input is essential for every step of the procedure for accurate record of the decision maker's preferences. This makes the method extremely data intensive. But the level of input and amount of data may not be available for every decision-making problem. The preferences of the decision makers also need to be precise, giving specific weights to each of the consequences. These require stronger assumptions at each level. This can be hard for precise application and can be relatively subjective. There are many common applications of multi-criteria decision analysis techniques MAUT that lean heavily on its major strength. This is its ability to take uncertainty into account.

(a) Main Features of MCDM

In general, there are two distinctive types of MCDM problems due to the different problems settings: one type having a finite number of alternative solutions and the other an infinite number of solutions. Normally in problems associated with selection other an infinite number of solutions. Normally in problems of selection and assessment, the number of alternative solutions is limited. In problems of design, an attribute may take any value in a range. Therefore the potential alternative solutions could be infinite. If this is the case, the problem is referred to as multiple objective optimization problems instead of multiple attribute decision problems. Our focus will be on the problems with a finite number of alternatives. A MCDM problem can be described using a decision matrix. Suppose there are m alternatives to be assessed based on n attributes, a decision matrix is a $n \times m$ matrix with each element y_{ij} being the j-th attribute value of the i-th alternative.

(i) Multiple attributes/criteria often form a hierarchy.

Almost any alternatives, such as an organization, an action plan or a product any kind, can be evaluated on the basis of attributes. An attribute is a property of alternatives in question. Some attributes may break down further into lower levels of attributes, called sub-attributes. To evaluate an alternative, a criterion is set up for each attribute. Because of the one to one correspondence between attribute and criterion, sometimes attributes are also referred to as criteria and used interchangeably in the MCDM context. MCDM itself can also be referred to as Multiple Attribute Decision Analysis (MADA) if there are a finite number of alternatives.

(ii) **Conflict among criteria**.

Multiple criteria usually conflict with one another. For example, in designing a car, the criteria of higher fuel economy might mean a reduced comfort rating due to the smaller passenger space.

(b) **MCDM Solutions**

MCDM problems may not have a conclusive or unique solution. Therefore different names are given to different solutions depending on the nature of the solutions. These are explained below:

(c) **Ideal solution**

All criteria in a MCDM problem may be of two categories. Criteria that are to be maximized are in the profit criteria category, although they may not necessarily be profit criteria. Similarly criteria that are to be minimized are in the cost criteria category. An ideal solution to a MCDM problem would maximize all profit criteria and minimize all cost criteria. Normally this solution is not obtainable. The question is what would be a best solution for the decision maker and how to obtain such a solution?

(d) **Non dominated solutions**

If an ideal solution is an obtainable, the decision maker may take help of non-dominated solutions. An alternative solution is dominated if there are other alternatives that are better than the solution on at least one attribute and as good as it on other attributes. An alternative is called non dominated if it is not dominated by any other alternatives.

(e) **Satisfying solutions**

Satisfying solutions are a reduced subset of the feasible solutions with each Alternative more than all the expected criteria. A satisfying solution may be a dominated solution. Whether a solution is satisfying depends on the level of the decision maker's expectation.

(f) **Preferred solutions**

A preferred solution is a non-dominated solution satisfying the decision maker's expectations.

4.2.1   Analytical Hierarchy Process (AHP)

4.2.1.1   Definition

AHP [1] is a multi-criteria decision making technique that can help express the general decision operation by decomposing a complicated problem into a multilevel hierarchical structure of objective, criteria and alternatives [2]. AHP performs pairwise comparisons to derive relative importance of the variable in each level of the hierarchy and/or appraises the alternatives in the lowest level of the hierarchy in order to make the best decision among alternatives. The AHP method is based on the three principles, showing in Table 4.2.

When logical hierarchy is constructed, the decision makers can select the alternatives by making pair-wise comparisons for each of the chosen criteria (This comparison may use concrete data from the alternatives or human judgments as a way to input subjacent information [3]). Figure 4.1 showing the hierarchy tree of MCDM.

Scale of relative importance

The comparison between two elements using AHP can be done in different ways [4]. However, the relative importance scale between two alternatives suggested by Saaty [5] is the most widely used. Attributing values that vary from 1 to 9, the scale determines the relative importance of an alternative when compared to another alternative, as we can see in Table 4.3.

4.2.1.2 Working Principle

Let $\{C_1, C_2, C_3, \ldots, C_n\}$ and $\{A_1, A_2, A_3, \ldots, A_m\}$ be set of criteria and alternatives respectively. The main procedure for the conventional AHP (Fig. 4.2) can be described in a series of steps. In first step construct a pairwise comparison matrix (pairwise comparison on n criteria)

$$C = [a_{hk}]_{n \times n} \tag{4.13}$$

where $a_{kk} = 1$, $a_{kh} = 1/a_{hk}$. In the first step all this a_{hk} collected from Table 4.2.

In second step calculate $C^{2r}, r \in N$ (Here also r denote the number of iteration).

Table 4.2 Table showing basic principle of AHP

1st	2nd	3rd
Structure of the model	Comparative judgment of the alternatives	Synthesis of the priorities

Fig. 4.1 Example of a hierarchy of tree/criteria/objectives

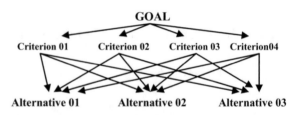

Table 4.3 Table showing scale of importance

Scale	Numerical rating	Reciprocal
Extremely preferred	9	1/9
Very strong to extremely	8	1/8
Very strongly preferred	7	1/7
Strongly to very strongly	6	1/6
Strongly preferred	5	1/5
Moderately to strongly	4	1/4
Moderately preferred	3	1/3
Equally to moderately	2	1/2
Equally preferred	1	1/1

In third step construct a normalized matrix

$$W_r = \left[w(r)_{ij}\right]_{n \times 1} \quad \text{from} \quad C^{2r} \tag{4.14}$$

where $w(r)_{ij} = b(r)_h / \sum_{h=1}^{n} b(r)_h$, $b(r)_h = \sum_{k=1}^{n} a_{hk} (h = 1, 2, \ldots, n)$ and $\sum_{i=1}^{n} w(r)_{ij} = 1$ (here j = 1).

In third step stop this iteration when absolute error of W_r and W_{r+1} is less than ϵ i.e.

$$W_r - W_{r+1} = \left[w(r)_{ij} - w(r+1)_{ij}\right]_{n \times 1} = [\epsilon]_{n \times 1} \tag{4.15}$$

where ϵ is very small positive real number. When termination condition satisfied then takes weightage of the criteria's was found from the matrix W_r and is denoted by W.

In step four calculate n number of matrix (each matrix of order $m \times m$) where these entire matrixes constructed by pair wise comparison of all alternatives $A_t(t = 1, 2, \ldots m)$ with respect to each criteria $C_j(j = 1, 2, \ldots n)$ and using above step calculate weightage of criteria from weightage vector (each matrix of order $m \times 1$). Combining all this weightage vector construct a matrix of order $m \times n$ and sum of each column is 1 and is denoted by A.

In step five calculate

$$A_{m \times n} W_{n \times 1} = \left[b_{ij}\right]_{m \times 1} \tag{4.16}$$

From $\left[b_{ij}\right]_{m \times 1}$ determined the weightage of alternatives.

4.2.1.3 Applications

AHP is widely applied in decision making problem solving various types of problems in almost all disciplines. In 2001 Robert and James suggest a new method

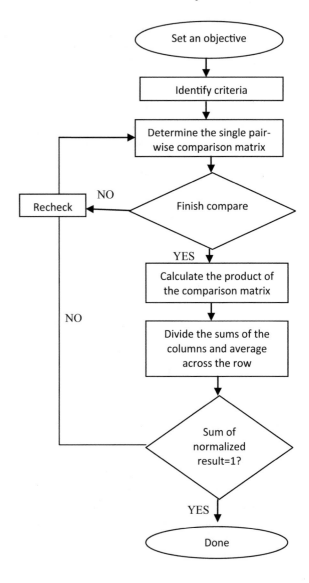

Fig. 4.2 Figure depicts the basic methodology of taking decision by AHP method [6]

of finding the fuzzy weights in fuzzy hierarchical analysis, which is the direct fuzzification of the λ_{max} method, used by Saaty in the analytical hierarchical process [7]. Ali [8] applies AHP on GIS-based landslide susceptibility mapping. Akash et al. [9] use AHP methodology to perform a comparison between the different electricity power production options in Jordan. Ta and Har [10] findings a survey of undergraduates on bank selection preferences. Robert and Ernest [11] using AHP for group decision making and offers numerous benefits as a synthesizing mechanism in group decisions. Some of applications of AHP given in the reference [12–16], where both qualitative and qualitative range are required to be included.

4.2.2 Fuzzy Decision Making

4.2.2.1 Definition

Zadeh [17] introduced the term fuzzy set and logic and described the Mathematics as fuzzy set theory. He [18] also made the distinction between fuzzy logic in broad sense (FLb) and fuzzy logic in narrow sense (FLn). Fuzzy sets represent linguistic level or term sets (or weights) such as slow, low, medium, high etc. Fuzzy membership function represents term sets, commonly used membership functions are triangular, trapezoidal, bell shaped Gaussian curve. In fuzzy logic the truth of any statement is a matter of degree. In fuzzy logic operators such as and, or and not are implemented by intersection, union and compliment operators.

Fuzzy logic in narrow sense: Fuzzy logic is a branch of fuzzy set theory, which deals with the representation and inference from knowledge. Fuzzy, unlike other logical systems, deals with imprecise or uncertain knowledge. In this narrow and perhaps correct sense, fuzzy logic is just one of the branches of fuzzy set theory.

Fuzzy logic in broad sense [19]: Flb is an extension of FLn, which aims developing a formal theory of human reasoning with the stress to the utilization of vagueness contain in the meaning of natural language expression.

4.2.2.2 Working Principle

In case of Fuzzy Logic the littoral rating is first converted to its crisp equivalent by 11point fuzzy scale (Table 4.4). The main procedure for the conventional Fuzzy Logic can be described in a series of steps. In first step construct a pairwise comparison matrix (pairwise comparison on n criteria)

$$D = [d_{hk}]_{n \times n} \tag{4.17}$$

Table 4.4 Table sowing 11-fuzzy scale

Ranking	Short form	Full form
1	EXLI	Exceptionally low important
2	ELI	Extremely low important
3	VLI	Very low important
4	LI	Low important
5	SLI	Semi low important
6	NHNLI	Neither high nor low important
7	SHI	Semi high important
8	HI	High important
9	VHI	Very high important
10	EHI	Extremely high important
11	EXHI	Exceptionally high important

where $d_{hk} = 12 - d_{kh}$.

In the second step all this a_{hk} collected from Table 4.4. In second step find $m_i = \max_j d_{ij}(\forall i = 1, 2, \ldots n)$. In third step construct a normalized matrix

$$E = \left[e_{ij} \right]_{n \times 1} \quad \text{from} \ D \tag{4.18}$$

where $e_{ij} = v_i / \sum_{i=1}^{n} v_i$ and $v_i = \sum_{j=1}^{n} d_{ij} / m_i$. Then from E weightage of the criteria's was found.

In step four calculate n number of matrix (each matrix of order $m \times m$) where these entire matrixes constructed by pair wise comparison of all alternatives $A_t(t = 1, 2, \ldots m)$ with respect to each criteria $C_j(j = 1, 2, \ldots n)$ and using above step calculate weightage of criteria from weightage vector (each matrix of order $m \times 1$). Combining all this weightage vector construct a matrix of order $m \times n$ and sum of each column is 1 and is denoted by F.

In step five calculate

$$F_{m \times n} E_{n \times 1} = \left[z_{ij} \right]_{m \times 1} \tag{4.19}$$

From $\left[z_{ij} \right]_{m \times 1}$ determined the weightage of alternatives.

Table 4.5 Table showing the location information of the large scale hydropower projects selected as alternatives for the present investigation

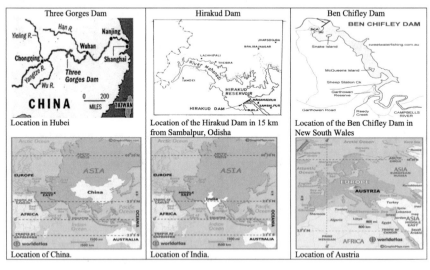

4.2.2.3 Applications

Fuzzy Logic is applied in widely decision making problem solving in various types of disciplines. Tanino [20] discussed Fuzzy decision making for fuzzy preference orderings in group decision making. In 1997 Sukha may use Fuzzy decision making for Min-transitivity of fuzzy leftness relationship and its application [21]. Olcer and Odabasi [22] use a new fuzzy multiple attribute decision-making (FMADM) method, which is suitable for multiple attributive group decision making (GDM) problems in fuzzy environment, is proposed to deal with the problem of ranking and selection of alternatives. Zeng et al. [23] use Fuzzy decision making for a risk assessment methodology to cope with risks in complicated construction situations. Some other application of Fuzzy Logic which are acclaimed widely in the ones the world giving in the reference [24–30].

4.2.3 Analytic Network Process (ANP)

4.2.3.1 Definitions

ANP was proposed by Saaty (1996) [31] to overcome the problem of dependence and feedback among criteria or alternatives [32]. Furthermore, the ANP method (Fig. 4.4) is used to decide the relative weights of the criteria. It improves the visibility of decision-making processes and generates priorities between the decision alternatives.

4.2.3.2 Working Principle

The main procedure for the conventional ANP can be described in a series of steps. In first step of ANP consists of the step 1, 2, 3 and 4 of AHP.

In second step calculate m number of matrix (each matrix of order $n \times n$) where these entire matrixes constructed by pair wise comparison of all criteria $C_t(t = 1, 2, \ldots n)$ with respect to each alternatives $A_j(j = 1, 2, \ldots m)$ and similar way using the step 1, 2 and 3 of AHP calculate weightage of alternatives from weightage vector (each matrix of order $n \times 1$). Combining all this weightage vector construct a matrix of order $n \times m$ and sum of each column is 1 and is denoted by G.

In third step In case of ANP construct a problem into a complete set of hierarchical or network model, then similar way of AHP generating pairwise comparisons to estimate the relative importance of various elements at each level, constructing a super matrix (showing in matrix 1; Where S is the super matrix since its elements are matrices, $W_{n \times 1}$ is a matrix that represents the impact of the criteria and $A_{m \times n}$ is a matrix that represents the impact of the alternatives on each of the criteria, $G_{n \times m}$ is a matrix that represents the impact of the criteria on each of the alternatives) to represent the influence priority of elements; after construction of Super matrix normalized it's values and the value of each cluster then this super matrix is called weighted super matrix; finally obtaining a large $K \in N$ such that S^{2K+1} allows convergence of the interdependent relationship this matrix is called decisions based on the super matrix from this matrix get the decision making result.

Matrix-1: Super-matrix

	Goal	Criteria	Alternatives
Goal	0	0	0
Criteria	$W_{n \times 1}$	0	$G_{n \times m}$
Alternatives	0	$A_{m \times n}$	0

4.2.3.3 Applications

ANP is applied in widely decision making problem solving in various types of disciplines. First time Meade and Presley [33] use ANP for discusses the requirements of the research and development project selection problem, which requires the allocation of resources to a set of competing and often disparate project proposals. Wolfslehner et al. [34] apply ANP for sustainable forest management. Hsu and Hu [35] use ANP approach to incorporate the issue of hazardous substance management (HSM) into supplier selection. Cheng et al. [36] employment of the ANP to select the best site for a shopping mall. Some other application of ANP which are acclaimed widely in the ones the world given in the references [37–41].

4.2.4 Fuzzy Analytic Network Process (FANP)

4.2.4.1 Definition

FANP proposed by Chang [42], Kahraman et al. [43], Mikhailov [44, 45]. A fuzzy logic is introduced in the pairwise comparison of ANP to make up for this deficiency in the conventional ANP, and is called as FANP [44].

4.2.4.2 Working Principle

Firstly, constructing pairwise matrices of the components with fuzzy judgments and determining the local priorities, consistency index from each matrix using the fuzzy prioritization method.

Secondly, checking the consistency index and aggregate local priorities into group priorities using nonlinear programming approach as explained.

Thirdly, filling the super matrix with the elicited group priorities to form unweighted super matrix, and obtain weighted super matrix by multiplying the unweighted super matrix by the corresponding cluster priorities, and then adjusting the resulting super matrix to column stochastic.

Fourthly, limit the weighted super matrix by raising it to sufficiently large power so that it converges into a stable super matrix (all columns being identical) and normalize the scores of alternatives from the limit weighted super matrix into final priorities. Through the above steps, we can obtain relative to the general objective factors, and ultimately the list weight of each factor. As using ANP method to solve practical problems is complex. Also Fig. 4.5 showing the computation process of FANP.

4.2.4.3   Applications

FANP is applied in widely decision making problem solving in various types of disciplines. Yuksel and Dagdeviren [46] Use the fuzzy ANP for Balanced Scorecard. Promentilla et al. [47] apply fuzzy ANP for multi-criteria evaluation of contaminated site remedial countermeasures. Liu and Lai [48] use fuzzy ANP for Decision-support for environmental impact assessment. Razmi et al. [49] use for designing a decision support system to evaluate and select suppliers. Some other application of FANP which are acclaimed widely in the ones the world showing in the references [50–53].

4.3   ANN

4.3.1   Definition

Artificial Neural Network (ANN) (Fig. 4.6) is a powerful, robust and adaptive mathematical model for functional approximation, pattern recognition and classification process.

An ANN consists of set of interconnected elements namely 'neurons' (Fig. 4.7). All this neurons (or inputs) can be expressed as a linear combination of each input multiplied by a coefficient known as 'weight' and then added a term with that linear combination which is called as 'bias'. After construction of linear combination it is applied to a function that is often non-linear, called the 'activation function', by which the neuron's output is calculated. So it is clear that in ANN an output from a series of inputs [54–56].

4.3.2 Selection of Topology

It depends wholely on problems to which the network is being applied, that how many number of input nodes, N and output nodes, M are there in an ANN. But no fixed rules are there for number of nodes to be included in hidden layer. There may be two problems which may arise regarding the number of nodes in hidden layers. Firstly if the numbers of nodes are few networks can face problems in generalising to problem it has encountered before. Lastly if the numbers of nodes are large then network may take long time in learning anything of any value.

Figure 4.3 provides a closer depiction of an individual neuron (in the hidden and output layers). Each neuron, **j**, has a number of input arcs **u₁** to **u_j**. Associated with

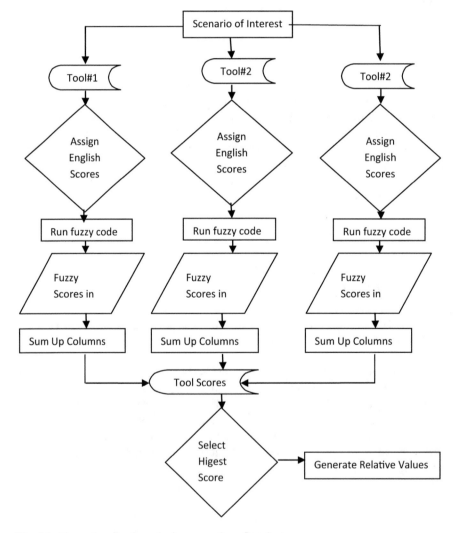

Fig. 4.3 Figure showing fuzzy logic process by a flowchart

each arc, **i**, is a weight, $\mathbf{w_{ij}}$, which represents a factor by which any values passing into the neuron are multiplied. A neuron, **j**, sums the values of all inputs according to Eq. (4.1):

$$S_j = \sum_{i=1}^{n} w_{ij}u_j + w_{0j} \qquad (4.20)$$

In Eq. (4.1) term, w_{0j} has been included called a bias. An activation function is applied to the value S_j, to provide the final output from the neuron. This activation function can be linear, discrete, or some other continuous distribution functions.

Application:

Kustrin and Beresford [59] apply ANN in pharmaceutical research. Kalogirou [60] use ANN for various applications of neural networks in energy problems in a thematic rather than a chronological or any other way. Okuyucu et al. [61] use ANN for the friction stir welding of aluminum plates. Vankayala and Rao [17] use ANN for a bibliographical survey of the research and explosive development of many ANN-related applications in electric power systems based on a subset of over 60 published articles [62].

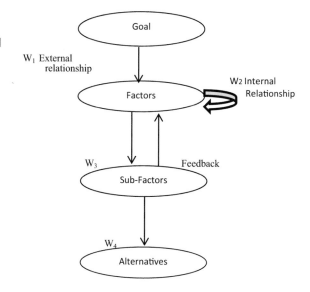

Fig. 4.4 Figure depicts the basic methodology of taking decision by ANP method [57]

Fig. 4.5 Figure depicts the
basic methodology of taking
decision by FANP method
[58]

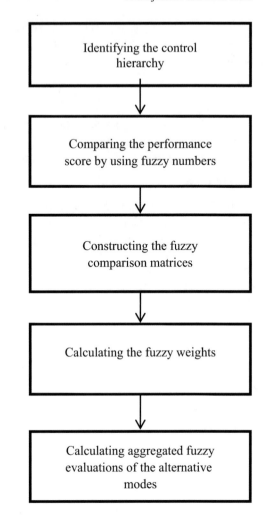

Fig. 4.6 A basic overview of
artificial neural network
topology

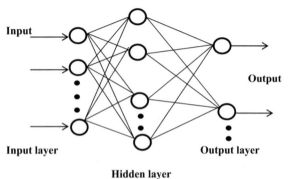

Fig. 4.7 An artificial neuron

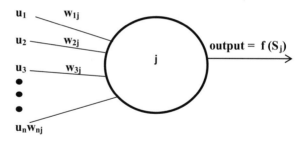

4.4   Methodology

There are three objectives in the present investigation. Among these three objectives, first objective is to find the most important cost factor which can induce maximum liability on the plant capital. The second objective is to find the least sensitive financial instrument HPP and finally the last objective of the study is to create a cognitive fool which can analyse the financial liability of any HPP based on the importance of each factor on the profitability of the proposed project and the status of the project with respect to the selected factors.

If f be the set which represents the financial factors of a project and d be the independent parameters which can determine the level of the project than we can say,

$$d \in f$$

and $f = g(d)$

where g is a function of real numbers.

$$d = \{costs\}$$

Now all the cost factors will not be uniformly influential. The importance of the factors will vary with respect to location and potential capacity.

Therefore $d = f(s, c)$

where s = location, c = potential capacity.

The present investigation try to find the importance of each factors with respect to location and capacity by conducting a literature Expert and stakeholders sure. The result of the survey is to find the weightages of importance of the identified parameters based on its influence in changing the cost factors of a plant.

In this regard MCDM techniques were utilized to induce an objective and the logical decision making. The MCDM requires three steps:

1. Selection of criteria
2. Selection of alternatives
3. Selection of Aggregation Method

Which are described next.

4.4.1 Application of the MCDM

In the present study the weights of importance of the identified cost parameters were determined with respect to AHP. Thus AHP, ANP, Fuzzy Decision Making and FANP method was adopted to find the weightage of importance for each of the cost parameters. The steps below provide the methodology adopted to determine the weightage of importance for each of the cost parameters by the MCDM methods.

4.4.1.1 Selection of Criteria

In the present study the weightage of importance of the cost parameters are required to be estimated. So all the cost parameters are considered as alternatives. To find the weightage of importance some criteria has to be identified with respect to which the alternatives will be compared and the difference in importance can be determined. In this regard the following factors are considered as Criteria:

(a) **Literature survey**:

A survey was carried out within literature of related fields from where it was found that the percentage of each cost parameters (Eq. 4.21) which can induce effect on the plant income. The score of literature survey is given by Eq. 4.21.
Let,

f   No of literature which prefer the alternative.
F   Total number literature studied.

$$C_1 = \text{Score of literature survey} = \frac{f}{F}\%$$ (4.21)

(b) **Expert survey**:

A survey was carried out within experts of related fields where participants were asked to suggest about the cost parameters which will be important for analysis of financial liability of certain projects.According to response received from the experts a percentage was given to the factors according to Eq. 4.22. Here Expert survey denoted by C_2.

Let, r = No. of expert survey which prefer the criteria.

R = Total number expert survey.

$$C_2 = \text{Score of expert survey} = \frac{r}{R}\%$$ (4.22)

(c) **Local survey**:

A survey was carried out within local people from an area where a HPP is already installed. Participants were asked to suggest about the cost parameters which can induce effect on the plant income. According to response received from the local people a percentage was given to the factors according to Eq. 4.23. Here Local survey denoted by C_3

Let, g = No of local survey which prefer the criteria.

G = Total number local survey.

$$C_3 = \text{Score from local survey} = \frac{g}{G}\% \qquad (4.23)$$

4.4.1.2 Selection of Alternatives

The alternatives were identified in the previous step. Various cost factors are considered as alternatives like Maintenance and operation cost (m) is $m = \left(\frac{u}{v} - \frac{a}{v(p \times o)}\right)$, Capital cost (c) is $c = \frac{d(1-(1+i)^{-n})}{i}$, Total production cost (T) is $T = p_v - r$, Project cost (j) is $j = \frac{\sum_l k_l}{L}(1-e)$, Energy equipment cost (E) is $= \left(\sum_{t=0}^{s} \frac{(I_t + m_t)}{(I_0 + R)^t}\right)\left(\sum_{t=0}^{s} \frac{F_t}{(I_0 + R)^t}\right)^{-1}$, Plant cost (y) is $y = f_1(\beta_1, \omega, \partial, \beta_2)$, Civil work cost (x) is $x = \sum_{hg} \sigma_{hg}$, Turbine cost (b) is $b = f_2(\alpha)$, Investment cost (β) is $\beta = f_2(\hbar, \varphi, \psi)$, Initial investment cost (δ) is $\delta = \frac{\mu - \tau}{\theta}$, Fuel cost (H) is $H = f_3(G)$, Labour cost (A) is $A = f_4(\sum s, \Gamma, \mathfrak{z})$ which are denoted by A_1, A_2, A_3, A_4, A_5, A_6, A_7, A_8, A_9, A_{10}, A_{11} and A_{12} respectively.

4.4.1.3 Selection of Aggregation Methods

In the present investigation AHP, ANP, Fuzzy Decision Making and FANP is utilized for identifying the weightage of importance of the parameter due to the availability of both quality and quantitative parameters.

In AHP putting $n = 3$ then from Eq. 4.13 is of the form $C = [a_{hk}]_{3\times3}$ and using Eq. 4.14 get $W_r = \left[w(r)_{ij}\right]_{3\times1}$, where $r = 12$, which gives the weightage of criteria matrix $W_{3\times1}$. At the end putting $m = 12$ (number of alternatives) in Eq. 4.16 get $A_{12\times3} W_{3\times1} = [b_{ij}]_{12\times1}$, which gives the weightage of alternatives matrix $[b_{ij}]_{12\times1}$.

In Fuzzy Decision Making putting $n = 3$ then from Eq. 4.17 is of the form $D = [d_{hk}]_{3\times3}$ and using Eq. 4.18 get $E = [e_{ij}]_{3\times1}$, which gives the weightage of criteria matrix $E_{3\times1}$. At the end putting $m = 12$ (number of alternatives) in Eq. 4.19

get $F_{m\times n}E_{n\times 1} = \left[z_{ij}\right]_{m\times 1}$, which gives the weightage of alternatives matrix $\left[z_{ij}\right]_{12\times 1}$. In case of ANP putting n = 3, m = 12 in matrix 1 get the super matrix.

S =	Super-matrix			
		Goal	Criteria	Alternatives
	Goal	0	0	0
	Criteria	$W_{3\times 1}$	0	$G_{3\times 12}$
	Alternatives	0	$A_{12\times 3}$	0

Where $W_{3\times 1}$ is a matrix that represents the impact of the criteria and $A_{12\times 3}$ is a matrix that represents the impact of the alternatives on each of the criteria, $G_{3\times 12}$ is a matrix that represents the impact of the criteria on each of the alternatives. At the end taking $\lim_{k\to\infty} s^k$, which gives the weightage of alternatives.

Similar way using the methods Fuzzy Decision Making and ANP get FANP.

4.4.1.4 Development of the Index

After the weightage of importance is determined an index was developed with the help of the weightage and the magnitude of the cost parameters. The weighted average of all the parameters is proposed as the index for $I \propto$ liability Index.

$$I = \frac{\sum_{i=1}^{n}\left(w_i \times A_i\right)}{\sum_{i=1}^{12} w_i}, \quad \text{where} \sum_{i=1}^{12} w_i = 1 \tag{4.24}$$

where w_i = weight of the alternatives, A_i = magnitude of the alternatives of any new HPP proposal.

4.4.2 Application of Neural Network and Genetic Algorithm

The objective of this study is to create a cognitive tool that can analyse the economic liability of any HPP based on the importance of each factor on the profitability of the proposed project In this regard some algorithms must be prepared so that I can be automatically calculated once the values of the A_i parameters are given as input. Due to the popularity of ANN, in mapping non-linearity and unknown relationships the said algorithm is applied to estimate the Liability Index (I) once the magnitude of the input parameters are entered in the framework. Another reason to apply ANN is to remove the requirement of repeated application of the MCDM methods once a new alternative is added. In the present study the ANN models were applied to predict the decision for the new alternative based on the existing knowledge that was gained from the available set of data.

(a) **Preparation of the Training Dataset**

The ANN models require training data so that the model can learn the inherent relationship between the input and output. In this aspect, the model requires to learn the inherent situations that may exist in reality and the way the output will represent the result. Thus, instead of applying real life data to train the ANN model, which will also make the model case wise acceptable, a normalized set of data depicting most of the situation that may arrive in the real life was utilized as training dataset. The training weights performed with various algorithms and the model which yielded better result is retained for prediction.

(b) **Topology Identification**

The number of hidden layers is responsible for quick learning of the problem but also increase the load on the computational infrastructure. That is why, selection of an optimal number of hidden layers is important for efficient performance of the neural network models and in the present study the said task was performed with the help of genetic algorithms where 50 generations were produced from 40 populations. The cross over rate was fixed at 0.8 whereas the mutation rate was controlled within 0.2 [63, 64].

(c) **Training the Network**

The ANN model was performed with three different algorithms, viz, Conjugate Gradient Descent, Quick Propagation and Levenberg Marquadart and Root Mean Square Error (RMSE), Relative Error (RE), Absolute Error (AE), Nash-Sutcliffe Efficiency (NSE) and Covariance (Var) between the predicted and actual output of the model is calculated to find the accurately trained network. The ANN model with selected training algorithm was used to predict I to the values of the A_i parameter provided to the modelling framework.

4.4.3 Validation of the Framework

The performance of the framework and the validation of the same was done by using sensitivity analysis and finding the Vulnerability Index of various places around the world The real situation were then compare with the index results.

4.4.3.1 Sensitivity Analysis

The sensitivity analysis was done by Multiple Input One output Tornado method developed by SensIt Limited. The range for the input variables were varied between 0 and 1. The impact of each input is then examined on the output and the results were compared with the weights of the variables found from the MCDM analysis.

4.4.3.2   Case Study

In this study three regions having a large- hydro, medium-hydro and Small-hydro power project. Three Gorges Dam (Table 4.5) is a high scale dam and it is denoted by L_1. Location of the Three Gorges Dam is in Hubei of China, which is also a new and largest dam in the world. The height of the dam is 181 m, length is 2,335 m and capacity is 39.3 km^3. Hirakud Dam is a medium scale dam and it is denoted by L_2. Location of the Hirakud Dam (Table 4.5) is in Sambalpur, Odisha of India, it is one of the oldest dam in the world, height of the dam is 60.96 m (200 ft), length is 4.8 km and capacity is 5.896 km^3. Type of the dam is Gravity dam. Ben Chifley Dam is a small scale dam and it is denoted by L_3. Location of the Ben Chifley Dam (Table 4.5) in New South Wales of Austria. The height of the dam is 34.4 m (113 ft), length is 455 m and capacity is 30,800 ml. Type of the dam is Embankment dam.

4.5   Results and Discussion

4.5.1   MCDM

According to the results from AHP decision making process, it was found that Maintenance and operation cost has the highest weightage of importance compared to the other 11 alternatives (0.16963 from Fig. 4.9). The lowest weightage of importance alternative was found for Labour cost (0.01787 from Fig. 4.9).

In AHP method literature review has the highest importance (0.46185 from Fig. 4.8) whereas among the criteria considered expert survey was found to have the lowest importance (0.25065 from Fig. 4.8) compared to other types of surveys considered in the present investigation.

Again with respect to the fuzzy decision making, ANP and FANP it was observed that the same alternative as selected by AHP have the highest weightage of importance (0.10218, 0.16925 and 10,263 from Fig. 4.9) compared to the other 11 options.

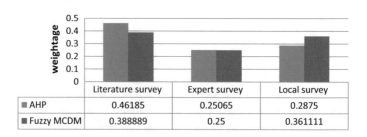

Fig. 4.8 Figure showing the weightage of importance of each criteria as determined by MCDM

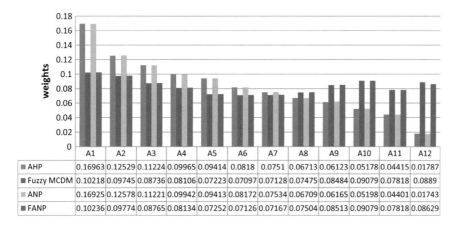

	A1	A2	A3	A4	A5	A6	A7	A8	A9	A10	A11	A12
■ AHP	0.16963	0.12529	0.11224	0.09965	0.09414	0.0818	0.0751	0.06713	0.06123	0.05178	0.04415	0.01787
■ Fuzzy MCDM	0.10218	0.09745	0.08736	0.08106	0.07223	0.07097	0.07128	0.07475	0.08484	0.09079	0.07818	0.0889
■ ANP	0.16925	0.12578	0.11221	0.09942	0.09413	0.08172	0.07534	0.06709	0.06165	0.05198	0.04401	0.01743
■ FANP	0.10236	0.09774	0.08765	0.08134	0.07252	0.07126	0.07167	0.07504	0.08513	0.09079	0.07818	0.08629

Fig. 4.9 Figure showing the weightage of importance of each alternative as determined by MCDM

Fig. 4.10 Showing values for the 3 cities with different country, as predicted by the neurogenetic model

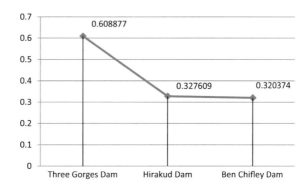

In Fuzzy decision making method also literature review has the highest importance (0.38889 from Fig. 4.8) whereas expert survey was found to have the lowest importance (0.25065 from Fig. 4.8) compared to other types of surveys considered to achieve the present objective.

The sensitivity of MCDM techniques can be adjusted by the impact of change in the importance of any one factor upon the overall ranking of all the factors. If the Rank 1 option remains unchanged then the MCDM is said to be insensitive and vice versa.

In the second test of sensitivity (If the first fails to the change in the ranking of all the alternatives identify the sensitiveness of the MCDM techniques) due to a minor change in the any one of the factors will depict the technique to be sensitive.

The level of sensitivity varies with the magnitude of change in any one factor which can result in the change in top ranked alternative or change in overall ranking of the factors (Table 4.6).

Table 4.6 Showing percentage of alternatives for difference criteria

Senerio		Percentage	A1	A2	A3	A4	A5	A6	A7	A8	A9	A10	A11	A12
1	A1	+10	+2.125877	−0.29152	−0.2608	−0.2341	−0.21796	−0.19571	−0.18475	−0.17349	−0.17159	−0.15973	−0.13652	−0.09879
2	A2	+10	+0.75595	+0.839451	−0.22465	−0.20187	−0.18769	−0.16903	−0.15992	−0.1508	−0.15005	−0.14064	−0.12025	−0.08958
3	A3	+10	+0.772932	−0.23841	+0.764014	−0.19163	−0.17822	−0.1604	−0.15168	−0.14291	−0.14202	−0.13292	−0.11364	−0.08418
4	A4	+10	+0.787425	−0.22787	−0.20387	+0.702212	−0.17034	−0.15331	−0.14497	−0.13658	−0.13572	−0.12702	−0.1086	−0.08043
5	A5	+10	+0.796817	−0.22046	−0.19723	−0.17711	+0.651257	−0.14815	−0.13997	−0.13164	−0.13049	−0.12178	−0.1041	−0.07621
6	A6	+10	+0.808442	−0.21247	−0.19009	−0.17076	−0.15883	+0.607153	−0.1351	−0.12723	−0.12635	−0.11817	−0.10103	−0.07463
7	A7	+10	+0.814055	−0.20874	−0.18676	−0.1678	−0.15603	−0.14049	+0.587307	−0.12528	−0.12459	−0.11672	−0.0998	−0.07424
8	A8	+10	+0.819473	−0.20551	−0.18387	−0.16531	−0.1536	−0.13851	−0.13117	+0.573689	−0.12363	−0.11623	−0.09939	−0.07502
9	A9	+10	+0.819243	−0.20691	−0.18514	−0.1666	−0.15461	−0.13979	−0.13265	−0.12579	+0.593343	−0.11928	−0.10204	−0.07884
10	A10	+10	+0.824724	−0.20393	−0.18248	−0.16435	−0.15235	−0.13807	−0.13123	−0.12484	−0.12574	+0.581936	−0.10224	−0.08051
11	A11	+10	+0.838276	−0.19325	−0.17292	−0.15565	−0.14439	−0.13065	−0.12403	−0.11774	−0.11823	−0.11197	+0.505931	−0.07448
12	A12	+10	+0.857005	−0.18177	−0.16267	−0.14672	−0.13574	−0.12352	−0.11775	−0.11264	−0.11431	−0.10953	−0.09377	+0.442337
13	A1	−10	−0.32545	+0.038507	+0.034502	+0.031628	+0.028653	+0.027244	+0.026805	+0.027079	+0.029474	+0.030286	+0.026029	+0.026164
14	A2	−10	+1.085929	−1.11959	+0.00385	−0.00256	−0.00347	−0.00106	+0.000472	+0.003038	+0.006657	+0.010079	+0.008809	+0.016467
15	A3	−10	+1.068189	−0.01762	−1.01277	−0.01325	−0.01335	−0.01006	−0.00812	−0.00519	−0.00171	+0.002042	+0.001927	+0.010849
16	A4	−10	+1.053107	−0.02858	−0.0255	−0.92354	−0.02154	−0.01744	−0.0151	−0.01177	−0.00826	−0.00409	−0.00332	+0.006964
17	A5	−10	+1.043386	−0.03625	−0.03237	−0.02832	−0.85709	−0.02277	−0.02027	−0.01688	−0.01367	−0.0095	−0.00796	+0.002608
18	A6	−10	+1.031355	−0.04451	−0.03976	−0.03489	−0.03345	−0.78973	−0.02531	−0.02144	−0.01795	−0.01323	−0.01114	+0.000979
19	A7	−10	+1.025567	−0.04836	−0.0432	−0.03794	−0.03634	−0.03069	−0.75831	−0.02346	−0.01976	−0.01473	−0.01241	+0.000573
20	A8	−10	+1.019997	−0.05168	−0.04616	−0.0405	−0.03884	−0.03273	−0.02936	−0.73237	−0.02075	−0.01524	−0.01282	+0.001374
21	A9	−10	+1.020269	−0.05021	−0.04484	−0.03915	−0.03778	−0.03139	−0.02782	−0.02291	−0.74843	−0.01207	−0.01008	+0.005328
22	A10	−10	+1.014703	−0.05322	−0.04752	−0.04142	−0.04007	−0.03313	−0.02924	−0.02385	−0.01852	−0.72402	−0.00985	+0.007067
23	A11	−10	+1.000793	−0.0642	−0.05735	−0.05037	−0.04824	−0.04075	−0.03664	−0.03116	−0.02626	−0.01958	−0.62614	+0.000835
24	A12	−10	+0.981942	−0.07572	−0.06764	−0.05932	−0.05692	−0.0479	−0.04293	−0.03625	−0.03014	−0.02195	−0.01846	−0.52377
25	A1	+20	+3.296842	−0.4489	−0.40163	−0.36085	−0.33556	−0.3021	−0.28574	−0.26931	−0.26776	−0.25074	−0.21438	−0.15895
26	A2	+20	+0.59653	+1.786886	−0.33138	−0.29822	−0.27673	−0.25024	−0.23747	−0.22522	−0.22588	−0.2136	−0.18273	−0.14101
27	A3	+20	+0.766307	−0.23169	−0.20737	+0.647508	−0.17368	−0.15295	−0.14209	−0.12968	−0.12267	−0.10848	−0.09256	−0.05173

(continued)

Table 4.6 (continued)

Senerio		Percentage	A1	A2	A3	A4	A5	A6	A7	A8	A9	A10	A11	A12
28	A4	+20	+0.658196	−0.32484	−0.29067	+1.493514	−0.24274	−0.21943	−0.20817	−0.19734	−0.19778	−0.18689	−0.15988	−0.12304
29	A5	+20	+0.676664	−0.31026	−0.27761	−0.24966	+1.386861	−0.20929	−0.19833	−0.18761	−0.18749	−0.17659	−0.15103	−0.11473
30	A6	+20	+0.699525	−0.29456	−0.26356	−0.23716	−0.22011	+1.289938	−0.18877	−0.17894	−0.17935	−0.16949	−0.14498	−0.11163
31	A7	+20	+0.71058	−0.2872	−0.25699	−0.23134	−0.21459	−0.1942	1.245679	−0.17509	−0.17588	−0.16662	−0.14256	−0.11085
32	A8	+20	+0.721268	−0.28083	−0.2513	−0.22641	−0.20979	−0.1903	−0.18102	+1.213101	−0.17399	−0.16565	−0.14176	−0.11239
33	A9	+20	+0.720847	−0.28356	−0.25377	−0.22895	−0.21176	−0.19282	−0.18391	−0.17609	+1.24949	−0.17166	−0.14697	−0.11992
34	A10	+20	+0.731726	−0.27764	−0.24849	−0.22446	−0.20727	−0.18937	−0.18108	−0.17419	−0.17809	+1.220308	−0.14733	−0.12319
35	A11	+20	+0.731726	−0.27764	−0.24849	−0.22446	−0.20727	−0.18937	−0.18108	−0.17419	−0.17809	+1.220308	−0.14733	−0.12319
36	A12	+20	+0.61233	−0.37882	−0.33916	−0.30783	−0.28248	−0.26148	−0.2524	−0.24697	−0.25823	+1.120326	−0.21898	+0.814617
37	A1	−20	−1.60948	+0.211678	+0.189452	+0.171029	+0.158066	+0.144168	+0.137697	+0.132119	+0.134647	+0.129543	+0.110932	+0.091076
38	A2	−20	+1.256755	−2.13268	+0.110395	+0.100549	+0.091854	+0.085835	+0.083426	+0.08258	+0.087646	+0.087935	+0.075476	+0.071155
39	A3	−20	+1.220476	+0.096213	−1.92835	+0.078699	+0.071648	+0.06743	+0.065865	+0.06576	+0.070545	+0.071522	+0.061423	+0.059696
40	A4	−20	+1.189697	+0.073847	+0.066168	−1.75879	+0.054928	+0.052376	+0.051625	+0.052337	+0.057196	+0.059021	+0.050735	+0.051783
41	A5	−20	+1.169914	+0.058243	+0.052194	+0.048002	−1.63047	+0.041532	+0.041103	+0.041947	+0.046205	+0.048025	+0.041298	+0.042932
42	A6	−20	+1.145431	+0.041427	+0.037159	+0.03462	+0.0307	−1.50431	+0.03086	+0.032666	+0.037497	+0.040427	+0.03483	+0.03962
43	A7	−20	+1.166646	+0.055745	+0.049999	+0.046041	+0.041467	+0.039872	−1.43537	−0.13664	−1.34681	+0.046533	+0.040049	+0.041953
44	A8	−20	+1.343698	+0.197363	+0.176788	+0.161637	+0.146916	+0.138721	−1.35833	+0.136127	−1.34681	+0.149508	+0.12843	+0.126877
45	A9	−20	+1.343698	+0.197363	+0.176788	+0.161637	+0.146916	+0.135118	−1.35833	+0.136127	−1.34681	+0.149508	+0.12843	+0.126877
46	A10	−20	+1.332133	+0.191119	+0.171223	+0.156922	+0.142185	+0.135118	−1.36068	+0.134184	+0.145967	−1.30671	+0.128931	+0.130532
47	A11	−20	+1.303095	+0.168196	+0.150696	+0.138239	+0.125103	+0.119185	−1.37305	+0.118908	+0.129799	+0.129002	−1.13057	+0.117489
48	A12	−20	+1.26411	+0.144394	+0.129445	+0.119745	+0.107174	+0.104431	−1.38343	+0.108435	+0.121835	+0.129002	+0.111036	−0.95525
49	A1	+30	+4.688162	−0.45256	−0.40486	−0.36311	−0.33845	−0.3032	−1.69554	−0.26741	−0.26324	−0.24369	−0.20821	−0.14697
50	A2	+30	+0.64124	+2.911795	−0.29949	−0.26915	−0.25021	−0.22541	−1.63762	−0.20125	−0.20038	−0.18793	−0.1607	−0.11998
51	A3	+30	+0.691467	−0.29721	+2.650259	−0.23885	−0.22219	−0.19987	−1.61811	−0.17788	−0.17658	−0.16507	−0.14112	−0.10393
52	A4	+30	+0.734498	−0.26591	−0.2379	+2.433268	−0.19879	−0.17879	−1.60216	−0.15906	−0.15785	−0.14751	−0.1261	−0.09276
53	A5	+30	+0.762525	−0.24377	−0.21808	−0.19563	+2.260683	−0.1634	−1.5902	−0.1443	−0.14222	−0.13186	−0.112267	−0.08013
54	A6	+30	+0.59225	−0.37541	−0.33593	−0.30256	−0.28046	+1.962597	−0.24163	−0.22989	−0.23157	−0.22005	−0.1883	−0.1481
55	A7	+30	+0.608582	−0.36455	−0.32622	−0.29397	−0.27231	−0.24716	+1.8947	−0.2242	−0.22645	−0.21582	−0.18472	−0.14696

(continued)

Table 4.6 (continued)

Senerio		Percentage	A1	A2	A3	A4	A5	A6	A7	A8	A9	A10	A11	A12
56	A8	+30	+0.624398	-0.35512	-0.31781	-0.28668	-0.2652	-0.24139	-0.23017	+1.843686	-0.22365	-0.21438	-0.18354	-0.14922
57	A9	+30	-0.28036	-0.23483	-0.21025	-0.19093	-0.17508	-0.16232	-0.15685	-0.15376	+1.986802	-0.15972	-0.137	-0.12477
58	A10	+30	-0.5346	-0.45072	-0.40357	-0.36693	-0.33595	-0.31245	-0.30261	-0.29783	+1.788406	+1.733703	-0.26821	-0.24832
59	A11	+30	-0.49849	-0.42023	-0.37627	-0.3421	-0.31323	-0.29129	-0.28212	-0.27764	+1.816093	-0.29136	+1.508937	-0.23138
60	A12	+30	-0.44797	-0.38559	-0.34534	-0.31512	-0.28715	-0.2697	-0.26302	-0.26203	+1.83194	-0.2835	-0.24344	+1.27186
61	A1	±30	-5.66215	+0.915573	+0.819428	+0.739759	+0.683644	+0.623592	+0.595599	+0.571596	+0.582654	+0.560821	+0.480257	-0.90985
62	A2	-30	+0.485875	-3.07191	+0.351642	+0.3173	+0.293417	+0.26729	+0.255047	+0.244331	+0.248462	+0.238526	+0.20423	+0.166714
63	A3	-30	+0.43312	+0.350334	-2.77647	+0.282928	+0.261616	+0.23835	+0.227451	+0.217931	+0.221663	+0.21285	+0.182249	+0.148901
64	A4	-30	+0.388541	+0.315132	+0.282041	-2.5328	+0.2353	+0.214672	+0.205065	+0.196851	+0.20073	+0.193279	+0.165517	+0.136598
65	A5	-30	+0.359799	+0.29081	+0.260262	+0.234822	-2.34564	+0.197784	+0.188688	+0.180697	+0.183663	+0.176227	+0.150885	+0.12293
66	A6	-30	+0.324585	+0.264289	+0.236548	+0.213719	+0.197302	-2.1668	+0.172541	+0.166071	+0.169948	+0.16427	+0.140705	+0.117742
67	A7	-30	+0.307738	+0.251934	+0.225504	+0.203944	+0.188032	+0.172359	-2.0848	+0.159596	+0.164116	+0.159454	+0.136619	+0.116428
68	A8	-30	+0.291694	+0.241215	+0.215936	+0.19565	+0.179951	+0.165783	+0.159487	-2.02147	+0.160864	+0.157728	+0.135208	+0.11888
69	A9	-30	+0.292958	+0.245619	+0.219914	+0.199748	+0.183124	+0.16985	+0.164191	+0.161048	-2.07806	+0.167563	+0.14373	+0.131234
70	A10	-30	+0.277222	+0.235965	+0.211308	+0.192445	+0.17581	+0.164258	+0.159602	+0.157976	+0.167629	-2.02234	+0.144378	+0.136671
71	A11	-30	+0.236445	+0.201408	+0.180363	+0.164284	+0.150058	+0.140248	+0.136306	+0.134978	+0.143304	+0.143883	-1.74746	+0.117104
72	A12	-30	+0.183194	+0.165192	+0.148027	+0.136137	+0.12278	+0.117783	+0.116536	+0.118992	+0.131106	+0.136411	+0.117307	-1.49254

4.5.2 Liability Index

As AHP was found to be better method than other MCDM the weightage of importance as selected by the former was taken as weightage for estimation of Liability Index (*I*) by Eq. 4.24. The liability index was created in such way that the value of the index will be directly proportional to liability of the project. The Ensemble weights from all the weights determined by the MCDM techniques were used along with the normalize value of the factors for the location of interest to determine the liability index (EQN) of each proposals.

4.5.3 Validation of the Model

After the index was developed an ANN infrastructure was created to provide a frame work for automatic estimation of the index simple feed forward supervised neural network was used. The inputs of the model was all the considered factors and the index was taken as output. The training and the testing error (MSE) was found to be 0.00081 and 0.00089 respectively which depicts a satisfactory accuracy of the model. The total weight of the network was 30 which categorize the network in lighter weights (Tables 4.7 and 4.8).

The model sensitivity analysis was also performed and the result confirmed that the model is in cohesion with study objective (Fig. 4.11). The liability of three locations having three types of plant capacity was also determined from the model and the results are shown in Fig. 4.10. From the results it can be concluded that the model correctly predicted the liability of three Gorges Dam to be highest and Ben Chifley Damto be lowest (Tables 4.9, 4.10 and 4.11).

Table 4.7 Table showing pair–wise comparison matrix of criteria

1	6	7
1/6	1	2
1/7	1/2	1

Table 4.8 Table showing comparison matrix with the help of 11-point fuzzy scale

0	NHNLI	SHI
NHNLI	0	ELI
SLI	EHI	0

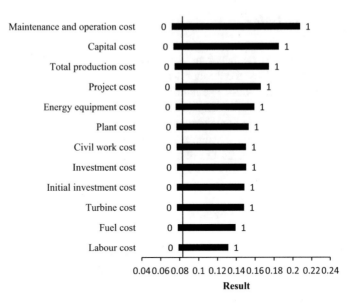

Fig. 4.11 Showing senivity for the model

Table 4.9 Showing the network preparation results

Network topology					
	Parameters	Results			
a.	Input	$A_1, A_2, A_3, A_4, A_5, A_6, A_7, A_8, A_9, A_{10}, A_{11}$ and A_{12}			
b.	Output	R			
c.	Topology	12 inputs	1 output	2 hidden layers with (2, 2) hidden units.	The network was trained on 385 records	Average absolute error: 0.02138
d.	Data subset selection	Training subset (68.14%)	Validation subset (15.93%)	Test subset (15.93%)		
e. The topology was created using genetic algorithms with following parameters						
1	Population size	40				
2	Number of generations	50				
3	Network size penalty	5				
4	Crossover rate	0.8				
5	Mutation rate	0.2				

Table 4.10 Showing the tanning results

Training results		
(a) Training subset		
1	Maximum absolute error	0.10987
2	Minimum absolute error	0.00007
3	Average absolute error	0.02138
4	Average MSE	0.00081
(b) Test subset		
1	Maximum absolute error	0.11594
2	Minimum absolute error	0.00023
3	Average absolute error	0.02082
4	Average MSE	0.00089

In the sensitivity analysis it is clearly indicated that Maintenance and Operation Cost is most and Labour cost has the least sensitivity towards the output. The weightages of importance as determined by MCDM also seconded this conclusion.

4.6   Conclusion

The present investigation tried to develop a new method for evaluation of the financial suitability of new hydro power projects. The study attempts to analyze financial liability of new hydro-power projects with the help of an index. The weights of importance of the factors were determined by multi-criteria decision making techniques after the important factors were identified through a literature, expert and stakeholders survey. An Artificial Neural Network model was also cascaded to the index so that values of the index for different situation can be automatically estimated without the requirement of re-accessing and recalculating the importance of the parameters and the index itself. The index was applied in three different regions with different levels of plant capacity. According to the results the index was successful to identify the projects as financially suitable. The different level of suitability as observed in the three selected cases was also identified by the variation in the magnitude of the index for the said places. The index can be now applied for the new project proposals of hydro power plant. One of the drawbacks of the index is accuracy of estimation depends largely on the amount of data sources surveyed and also on the techniques applied to determine the weights of importance. Although sensitivity analysis confirmed that the weights of importance assigned to different parameters represents their influence on the output successfully. Another limitation of the index is in its value which is a normalized representation of financial suitability. This value is made proportional with the financial suitability but does not exactly represent the financial liability. The correlation between the exact value of financial liability that can be incurred and the value of the index can be determined in future studies.

Table 4.11 Table showing normalized data for ANN

	A_1	A_2	A_3	A_4	A_5	A_6	A_7	A_8	A_9	A_{10}	A_{11}	A_{12}
L_1	0.969803	0.971581	0.969803	0.992309	0.5	0.5	0.892171	0.444444	0.428571	0.473684	0.470588	0.3125
L_2	0.016867	0.018009	0.016867	0.005514	0.333333	0.3125	0.058264	0.333333	0.333333	0.368421	0.352941	0.3125
L_3	0.013329	0.01041	0.013329	0.002177	0.166667	0.1875	0.049565	0.222222	0.238095	0.157895	0.176471	0.375

References

1. Saaty, T. L. (1980). The analytic hierarchy process: Planning, priority setting, resource allocation. New York: McGraw-Hill, Inc.
2. Sharma, M. J., Moon, I., & Bae, H. (2008). Analytic hierarchy process to assess and optimize distribution network'. *Applied Mathematics and Computation, 202,* 256–265.
3. Saaty, T. L. (2008). *Relative measurement and its generalization in decision making: Why pairwise comparisons are central in mathematics for the measurement of intangible factors— The analytic hierarchy/network process.* Madrid: Review of the Royal Spanish Academy of Sciences, Series A, Mathematics.
4. Triantaphllou, E., & Mann, S. H. (1995). Using the analytic hierarchy process for decision making in engineering applications: Some challenges. *International Journal of Industrial Engineering: Applications and Practice, 2*(1), 35–44.
5. Saaty, T. L. (2005). *Theory and applications of the analytic network process: Decision making with benefits, opportunities, costs, and risks.* Pittsburgh: RWS Publications.
6. Goh, H. H., Kok, B. C., Yeo, H. T., Lee, S. W., & Zin, A. A. M. (2013). Electrical power and energy systems. *International Journal of Electrical Power & Energy System, 47,* 198–204.
7. Csutora, R., & Buckley, J. J. (2001). Fuzzy hierarchical analysis: The Lambda-Max method. *Fuzzy Sets and Systems, 120,* 181–195.
8. Yalcin, A. (2008). GIS-based landslide susceptibility mapping using analytical hierarchy process and bivariate statistics in Ardesen (Turkey): Comparisons of results and confirmations. *CATENA, 72,* 1–12.
9. Akash, B. A., Mamlook, R., & Mohsen, M. S. (1999). Multi-criteria selection of electric power plants using analytical hierarchy process. *Electric Power Systems Research, 52,* 29–35.
10. Ta, H. P., & Har, K. Y. (2000). A study of bank selection decisions in Singapore using the analytical hierarchy process. *International Journal of Bank Marketing, 18,* 170–180.
11. Robert, F. D., & Forman, E. H. (1992). Group decision support with the analytic hierarchy process. *Decision support systems, 8*(2), 99–124.
12. Bevilacqua, M., & Braglia, M. (2000). The analytic hierarchy process applied to maintenance strategy selection. *Reliability Engineering & System Safety, 70,* 71–83.
13. Ghodsypour, S. H., & O'Brien, C. (1998). A decision support system for supplier selection using an integrated analytic hierarchy process and linear programming. *International Journal of Production Economics, 56–57,* 199–212.
14. Kurttila, M., Pesonen, M., Kangas, J., & Kajanus, M. (2000). Utilizing the analytic hierarchy process (AHP) in SWOT analysis—A hybrid method and its application to a forest-certification case. *Forest Policy and Economics, 1,* 41–52.
15. Liu, F. H. F., & Hai, H. L. (2005). The voting analytic hierarchy process method for selecting supplier. *International Journal of Production Economics, 97,* 308–317.
16. Ramanathan, R. (2001). A note on the use of the analytic hierarchy process for environmental impact assessment. *Journal of Environmental Management, 63,* 27–35.
17. Vankayala, V. S. S., & Rao, N. R. (1993). Artificial neural networks and their applications to power systems—A bibliographical survey. *Electric Power Systems Research, 28,* 67–79.
18. Zahir, S. (1999). Clusters in group: Decision making in the vector space formulation of the analytic hierarchy process. *European Journal of Operational Research, 112,* 620–634.
19. Novak, V. (2008). Principal fuzzy type theories for fuzzy logic in broader sense. Institute for Research and Applications of Fuzzy Modeling.
20. Tanino, T. (1984). Fuzzy preference orderings in group decision making. *Fuzzy Sets and Systems, 12,* 117–131.
21. Kundu, S. (1997). Min-transitivity of fuzzy leftness relationship and its application to decision making. *Fuzzy Sets and Systems, 86*(3), 357–367.
22. Olcer, A. I., & Odabasi, A. Y. (2005). A new fuzzy multiple attributive group decision making methodology and its application to propulsion/manoeuvring system selection problem. *European Journal of Operational Research, 2,* 93–114.

23. Zeng, J., An, M., & Smith, N. J. (2007). Application of a fuzzy based decision making methodology to construction project risk assessment. *International Journal of Project Management, 25,* 589–600.
24. Barreto-Neto, A. L. A., & Filho, C. R. S. (2008). Application of fuzzy logic to the evaluation of runoff in a tropical watershed. *Environmental Modelling and Software, 23,* 244–253.
25. Chen, S. M., & Tan, J. M. (1994). Handling multicriteria fuzzy decision-making problems based on vague set theory. *Fuzzy Sets and Systems, 67,* 163–172.
26. Gokmena, G., Akincib, T. C., Tektac, M., Onatc, N., Kocyigita, G., & Tekta, N. (2010). Evaluation of student performance in laboratory applications using fuzzy logic. *Procedia Social and Behavioral Sciences, 2,* 902–909.
27. Hong, D. H., & Choi, C. H. (2000). Multicriteria fuzzy decision-making problems based on vague set theory. *Fuzzy Sets and Systems, 114,* 103–113.
28. Karanovica, G., & Gjosevskab, B. (2012). Application of fuzzy logic in determining cost of capital for the capital budgeting process. *Procedia Economics and Finance, 3,* 78–83.
29. Lin, L., Yuan, X. H., & Xia, Z. (2007). Multicriteria fuzzy decision-making methods based on intuitionistic fuzzy sets. *Journal of Computer and System Sciences, 73,* 84–88.
30. Maji, P. K., Roy, A. R., & Biswas, R. (2002). An application of soft sets in a decision making problem. *Computers & Mathematics with Applications, 44,* 1077–1083.
31. Saaty, T. L. (1996). *'Decision making with dependence and feedback: The analytic network process.* Pittsburgh, PA: RWS Publications.
32. Liou, J. J. H., Tzeng, G. H., & Chang, H. C. (2007). Airline safety measurement using a hybrid model. *Air Transport Management, 13,* 243–249.
33. Meade, L. M., & Presley, A. (2002). R&D project selection using the analytic network process. *Engineering Management, 49,* 59–66.
34. Wolfslehner, B., Vacik, H., & Lexer, M. J. (2005). Application of the Analytic Network Process in multi-criteria analysis of sustainable forest management. *Forest Ecology and Management, 207,* 157–170.
35. Hsu, C. W., & Hu, A. H. (2009). Applying hazardous substance management to supplier selection using analytic network process. *Journal of Cleaner Production, 17,* 255–264.
36. Cheng, E. W. L., Li, H., & Yu, L. (2005). The analytic network process (ANP) approach to location selection: A shopping mall illustration. *Construction Innovation: Information, Process, Management, 5,* 83–97.
37. Bayazit, O. (2006). Use of analytic network process in vendor selection decisions. *Benchmarking: An International Journal, 13,* 566–579.
38. Lee, J. W., & Kim, S. H. (2000). Using analytic network process and goal programming for interdependent information system project selection. *Computers & Operations Research, 27,* 367–382.
39. Karsak, E. E., Sozer, S., & Alptekin, S. E. (2003). Product planning in quality function deployment using a combined analytic network process and goal programming approach. *Computers & Industrial Engineering, 44,* 171–190.
40. Wu, W. W., & Lee, Y. T. (2007). Selecting knowledge management strategies by using the analytic network process. *Expert Systems with Applications, 32,* 841–847.
41. Yuksel, I., & Dagdeviren, M. (2007). Using the analytic network process (ANP) in a SWOT analysis—A case study for a textile firm. *Information Sciences, 177,* 3364–3382.
42. Chang, D. Y. (1996). Applications of the extent analysis method on fuzzy AHP. *European Journal of Operational Research, 95,* 649–655.
43. Kahraman, C., Ertay, T., & Buyukozkan, G. (2006). A fuzzy optimization model for QFD planning process using analytic network approach. *European Journal of Operational Research, 171,* 390–411.
44. Mikhailov, L. (2003). Deriving priorities from fuzzy pairwise comparison judgments. *Fuzzy Sets and Systems, 134,* 365–385.
45. Mikhailov, L. (2004). A fuzzy approach to deriving priorities from interval pairwise comparison judgments. *European Journal of Operational Research, 159,* 687–704.

46. Yuksel, I., & Dagdeviren, M. (2010). Using the fuzzy analytic network process (ANP) for Balanced Scorecard (BSC): A case study for a manufacturing firm. *Expert Systems with Applications, 37,* 1270–1278.
47. Promentilla, M. A. B., et al. (2008). A fuzzy analytic network process for multi-criteria evaluation of contaminated site remedial countermeasures. *Journal of Environmental ManagementVol., 88,* 479–495.
48. Liu, K. F. R., & Lai, J. H. (2009). Decision-support for environmental impact assessment: A hybrid approach using fuzzy logic and fuzzy analytic network process. *Expert Systems with Applications, 36,* 5119–5136.
49. Razmi, J., Rafiei, H., & Hashemi, M. (2009). Designing a decision support system to evaluate and select suppliers using fuzzy analytic network process. *Computers & Industrial Engineering, 57,* 1282–1290.
50. Ayag, Z., & Ozdemir, R. G. (2011). An intelligent approach to machine tool selection through fuzzy analytic network process. *Journal of Intelligent Manufacturing, 22,* 163–177.
51. Dagdeviren, M., Yüksel, J., & Kurt, M. (2008). A fuzzy analytic network process (ANP) model to identify faulty behavior risk (FBR) in work system. *Safety Science, 46,* 771–783.
52. Dagdeviren, M., & Yuksel, J. (2010). A fuzzy analytic network process (ANP) model for measurement of the sectoral competitition level (SCL). *Expert Systems with Applications, 37,* 1005–1014.
53. Vinodh, S., AneshRamiya, R., & Gautham, S. G. (2011). Application of fuzzy analytic network process for supplier selection in a manufacturing organisation. *Expert Systems with Applications, 38,* 272–280.
54. MokhatabRafiei, F., Manzari, S. M., & Bostanian, S. (2011). Financial health prediction models using artificial neural networks, genetic algorithm and multivariate discriminant analysis: Iranian evidence. *Expert Systems with Applications, 38,* 10210–10217.
55. Poddig, T., & Rehkugler, H. (1996). A 'world' model of integrated financial markets using artificial neural networks. *Neurocomputing, 10,* 251–273.
56. Thawornwong, S., & Enke, D. (2004). The adaptive selection of financial and economic variables for use with artificial neural networks. *Neurocomputing, 56,* 205–232.
57. Sevkli, M., AsilOztekin, A., Uysal, O., Torlak, G., Turkyilmaz, A., & Delen, D. (2012). Development of a fuzzy ANP based SWOT analysis for the airline industry in Turkey. *Expert Systems with Applications, 39,* 14–24.
58. Tuzkaya, U. R., & Onut, S. (2008). A fuzzy analytic network process based approach to transportation-mode selection between Turkey and Germany: A case study. *Information Sciences, 178,* 3133–3146.
59. Kustrin, S. A., & Beresford, R. (2000). Basic concepts of artificial neural network (ANN) modeling and its application in pharmaceutical research. *Journal of Pharmaceutical and Biomedical Analysis, 22,* 717–727.
60. Kalogirou, S. A. (2000). Applications of artificial neural-networks for energy systems. *Applied Energy, 67,* 17–35.
61. Okuyucu, H., Kurt, A., & Arcaklioglu, E. (2007). Artificial neural network application to the friction stir welding of aluminum plates. *Materials and Design, 28,* 78–84.
62. Zadeh, L. A. (1975). Fuzzy logic and approximate reasoning. *Synthese, 30,* 407–428.
63. Holland, J. H. (1992). Genetic algorithm computer programs that "evolve" in ways that resemble natural selection can solve complex problems even their cantors do not fully understand. Scientific American.
64. Kherfane, R. L., Younes, M., Kherfane, N., & Khodja, F. (2014). Solving economic dispatch problem using hybrid GA-MGA. *Energy Procedia, 50,* 937–944.

Chapter 5
Wave Energy Potential Site Selection Based on MCDM and Neural Network Analysis

Soumya Ghosh

Abstract The present study, an improved wave energy potential estimate has been made. Based on various parameters such as physical site characteristics, environmental conditions and socio-economic regional state, the selection criteria have been suggested. This would form the basis for energy device selection for the decision makers. If analytical network process (ANP) is used to determine the weight vector to be assigned to the criteria considered for a certain decision-making problem, the output of the result will be more logical and the haziness of the conversion to a crisp rating will not influence the decision. Thus, we investigated a hybrid ANP method to identify the most suitable location for a wave energy potential site. The index also provided a heuristic and cognitive optimal value to way from a suitability of small scale hydro power plant installation. Both models were able to fit the data well, with R^2 values of 0.98462 and 0.9964 for the linear regression model and the ANN model respectively. According to the results, wave height was found to have maximum importance followed by wind speed, wave period, water depth and salinity. The total three different neural networks were developed to predict the same output, all the models of five input to have a optimal performance.

5.1 Introduction

Global warming and the greenhouse effect are the main reasons to investigate and incorporate clean fuel technologies and new energy sources around the globe [1–3]. These environmental issues and also the rapid depletion of fossil fuels encouraged the countries and different organizations to attain highly efficient and green power plants [4, 5]. Technological advancement helps in achieving some means of harvesting energy from the renewable sources and to use them as the source of new, clean and sustainable energy to meet the world's demand [6–10]. Renewable

S. Ghosh (✉)
School of Hydro-informatics Engineering, National Institute of Technology Agartala, Agartala, Tripura (W) 799046, India
e-mail: soumyaee@gmail.com

© Springer Nature Singapore Pte Ltd. 2018   107
M. Majumder (ed.), *Application of Geographical Information Systems and Soft Computation Techniques in Water and Water Based Renewable Energy Problems*, Water Resources Development and Management, https://doi.org/10.1007/978-981-10-6205-6_5

energy resources are regenerative and do not deplete over time. Renewable energy ensures improved energy security of the countries all over the world and reduces carbon emissions.

In nature, there are different kinds of renewable resources that potentially can be used for the production of clean energy. As the sun heats the earth, winds are generated to convey energy to the ocean surface in the form of wind-waves. Waves broadcast this stored energy thousands of kilometers without significant loss and hence wave energy becomes one of the most important renewable energy resources with low emission citation. In addition to its abundant solar, wind and geothermal resources is also uniquely situated to capture the renewable energy of the ocean. Special buoys, turbines, and other technologies can capture the power of waves and tides and convert it into clean, pollution-free electricity. The world energy demand is steadily increasing along with the population growth and improved living conditions. The energy from ocean waves is the most conspicuous form of ocean energy, possibly because of the, often spectacular, wave destructive effects. Wave energy is a renewable energy source with high power density, relatively high utilization factor, low visual impact and presumed low impact on environment compared to other renewable sources. Like other renewable resources, the wave energy is variable in nature. Waves are produced by winds blowing across the surface of the ocean. However, because waves travel across the ocean, their arrival time at the wave power facility may be more predictable than wind. In a maritime country with long coastlines, wave energy can potentially be harvested to meet the energy demands and reduce dependency on fossil fuel [11]. Many renewable resources like solar, wind, and ocean energy (tide and wave) are assumed as a proper alternatives for traditional energy sources [12]. The wave power resources in the world are estimated around 2 TW [13].

The total wave power was found to range from 1 GW in Sweden [9], to 120 GW in UK [14], passing through 3.4 GW of Denmark [15], 10 GW in Portugal [16], 21 GW in Ireland [17] and over 28 GW in the area of Gulf of Gascoigne, France [18], Spain, France, Italy and Greece is estimated to be 30 GW [19]. The eave energy resource of the US Pacific Northwest (i.e. off the coast of Washington, Oregon and N. California) has an average significant wave height 2 and 5 m and energy periods between 8 and 12 s to contribute the little annual energy. The mean wave power is 36 kW/m annually, 64 kW/m during the winter months and 12 kW/m in the summer months [20]. The area around Cape Estaca de Bares (the northmost position of Iberia) presents an excellent potential for wave energy exploitation due to its suitable position, with average deepwater wave power values exceeding 40 kW/m. The significant wave height between 2 and 5 m and during the energy periods between 11 and 14 s and wave direction was 300°. The Estaca de Bares area (Spain) presents a wave energy resource, with offshore average wave power and annual wave energy exceeding 40 kW m^{-1} and 350 MWh m^{-1} [21]. The annual wave energy output and power the Pelamis was highest from the Irish location followed by USA WC, Canada EC, Portugal, Canada WC and finally USA WC. The wave energy period recording the highest frequency of hours was 5–7 s in all locations. The wave height of approximately 1–2.5 m was most common for all locations, except Ireland, where the highest prevalence of wave height occurred between 1–3.5 m. the total yearly energy output formed by the Pelamis wave energy converter device was 2.5 GWh [22]. The Baltic Sea is the world's largest brackish water sea (~ 268.000 km^2), being relatively shallow with an average depth of only 55 m and with a maximum depth of 450 m. The annual average for waves is estimated to 5 kW/m and 0.7 kW/m^2 for wind and the energy flux in the Baltic Sea is higher than that of offshore wind power [23]. The island of El Hierro (Spain), a UNESCO Biosphere Reserve in the Atlantic Ocean, aims to become the first 100% renewable energy island in the world. The sea states with a significant wave height above 3 m in winter and autumn are rare in spring and very rare in summer and energy period ranges 6–16 s is selected. Therefore resource is found west and north of El Hierro, with average wave power in the order of 25 kW m^{-1} and total annual energy in excess 200 MWh m^{-1} [24].

The objective of the present study is to develop a method to select a suitable location for wave energy potential with the help of ANP method and GMDH. This paper focuses on the assessment of wave power potential in USA and Japan scenario.

5.2 Multi Criteria Decision Making

The Multi criterion Decision-Making (MCDM) are gaining importance as potential tools for analyzing complex real problems due to their inherent ability to judge different alternatives (Choice, strategy, policy, scenario can also be used synonymously) on various criteria for possible selection of the best/suitable alternative (s). These alternatives may be further explored in-depth for their final implementation.

5.2.1 Analytical Network Process (ANP)

The Analytic Network Process (ANP), first proposed by Saaty in 1996, is derived from the Analytic Hierarchy Process (AHP) [25]. AHP is one of the most popular methods used in multi-criteria decision analysis [26]. It decomposes a decision problem from the top overall goal to a set of manageable clusters, sub-clusters, and so on, down to the bottom level, which usually contains scenarios or alternatives [27]. Although both the AHP and the ANP derive ratio scale priorities by making paired comparisons of elements on a criterion, there are some differences between them. The first difference is that the AHP is a special case of the ANP, because the ANP handles dependence within a cluster (inner dependence) and among different clusters (outer dependence). Second, the ANP is a nonlinear structure, while the AHP is hierarchical and linear, with a goal at the top level and the alternatives on the bottom level [28].

There are four basic steps when using ANP: (1) deconstructing a problem into a complete set of hierarchical or network model; (2) generating pair wise comparisons to estimate the relative importance of various elements at each level; (3) building a super matrix to represent the influence priority of elements; and, (4) making decisions based on the super matrix [29]. The ANP methodology is considered an ideal choice for formulating the selection problem for several reasons. Firstly, the ANP is suitable for solving complex decision problems that are multi-criteria in nature. Often, real-life decision problems are rather complex and as such there is the need to take into account trade-offs between both tangible and intangible decision criteria [30]. To account for these trade-offs, the ANP methodology allows decision makers to express their preference between decision elements through the reciprocal pairwise comparison process. This comparison is based on the Saaty's fundamental scale [31]. Moreover, consistency in the decision making process is evaluated through computing the Consistency Ratio (CR).

5.3   Mechanism of Group Method Data Handling (GMDH)

The GMDH method is a self-adaptive heuristic ANN-based method proposed in the 1960s by Alexey Grigorevich Ivakhnenko in 1971. The original GMDH was a supervised inductive algorithm for construction of self-organizing models of optimal convolution based exclusively on the input–output relationships of a given dataset, without the need for user interference. The GMDH network is known as a self-organized approach that solves various complex problems in non-linear systems [32].

In study, the proposed cluster algorithm is build based on the theory of GMDH. The mechanism of GMDH is introduced in this section. The three components must be existing to maintained: initial solution, a transfer function and external criteria.

GMDH is an evolutionary approach which including operations of mutation and selection. It starts from the initial model set, carries on parameter estimation by a transfer function and obtains middle candidate solutions (inherit, mutation), evaluates the middle candidate solutions by external criterion [33], and finally chooses some best ones (and thus ignores the others) for the next layer. Therefore, the modeling process of the GMDH can be described as the following four steps:

Firstly, by using some initial solutions as inputs, GMDH prepares to generate middle candidate solutions in the first layer. Secondly, each pair of initial solutions is combined to generate the middle candidate solutions in the intermediate layers (inheritance, mutation). The transfer function is utilized to generate a set of candidate solutions. It describes the relationships between the initial solutions and the middle candidate solutions. Generally, the first order (linear) Kolmogorov–Gabor polynomial including n nodes can be used as transfer function [34]:

$$Y = f(x_1, x_2, \ldots, x_n) = a_0 + a_1 x_1 + a_2 x_2 + \cdots + a_n x_n \tag{5.1}$$

where Y is the middle candidate solution, x is a given initial solutions and a is the vector of coefficients or weights. New middle candidate solutions can be obtained according to the inputs of the current layer and the transfer function.

Thirdly, select some of the best candidate solutions as the inputs of the next layer. In GMDH, the selection of the middle candidate solutions is based on the value of a given external criterion on the testing dataset (selection). Then these selected candidate solutions are passed on to the next layer.

Finally, the above generation and selection process of the middle candidate solutions will repeat until an optimal complexity model found by the principle of termination. The principle of termination is presented by the theory of optimal complexity: along with the increase of model complexity, the value of the external criterion will decrease first and then increase, so the global extreme value corresponds to the optimal complexity model [35]. In this way, GMDH can determine the inputs and structure of the final model automatically. It is worthy to note that the number of iterations when build GMDH model is not a preset number, but determined by the optimal complexity principle which associates with the three components of the GMDH model.

5.4   Methodology

The five most important parameters for the efficiency of a wave energy potential were selected from are view of the literature. There are factors were wave height, wind speed, water depth, wave period and salinity. The present study applied a two steps developed an indicator for the potential performance with respect to the location based and efficiency based on the energy potential. The first step the literature survey of wave energy potential in global scenario were used their importance of the study objective. Next step MCDM method of ANP was applied to the priority of the selected parameters. The model uses ANP to determine weights of importance as derived from the rank of importance and the aggregation method. In Fig. 5.1 shows a schematic diagram of the model methodology.

5.4.1   Application of MCDM

The application of MCDM involves three steps.

5.4.1.1   Criteria

The criteria was selected based on study objective. The present study the location and efficiency were selected the criteria.

5.4.1.2   Alternative

The each of the alternatives was compared with each other based on the rank from the literature survey and data availability based method.

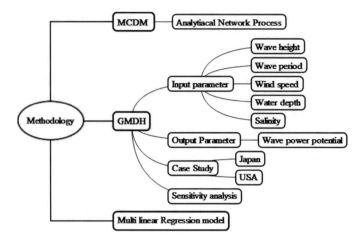

Fig. 5.1 Figure showing a methodology of the present investigation

5.4.1.3 Aggregation Method

The criteria and alternatives were used to find the equivalent weight of the selected parameters after each of the alternatives were compared with each other, based on the criteria and with respect to the study objective.

Both results were cross multiplied to get the weight of importance for each of the parameters based on the criteria and with respect to the study goal.

At the end of this method the importance of the parameters was represented by the weight vector or priority values of the parameters which are directly proportional to the significance of the variables.

Weight Function as the Indicator: The W-value or the Indicator for performance evaluation of array design of wave energy converter was estimated by Eq. 5.2

$$W_{value} = \frac{\sum w_n b_n}{\sum w_m b_m} \tag{5.2}$$

where w_n and b_n are, respectively, the magnitude of weight of importance, beneficiary and non-beneficiary variables and n and m are the number of beneficiary and non-beneficiary variables, respectively.

5.4.2 GMDH Model

The GMDH model was developed with twenty-one inputs and one output. The dataset of the models was normalized and then 60% of the data was used as training and 40% was kept for testing. In total, twelve models were developed with different combinations of the control charts, MCDMs and data transformations.

The performance of all the forty-eight models were analyzed by Mean Absolute Error (MAE) [29], Root Mean Square Error (RMSE) [36], Mean Relative Error (MRE) [37] and Correlation (R) [38]. The former metrics are known to be inversely proportional to model accuracy whereas the other metrics are directly proportional to model performance. The performance of the model during the checking (c) or testing phase is a more important indicator of model reliability than the performance of the model in the training (t) phase [39].

The performance of the selected three models were tested for reliability with the help of Root Mean Square Error (RMSE), Mean Relative Error (MRE), Percent bias (PBIAS) [40] between the predicted and observed data. The Performance Index (PI) was prepared to represent the performance of the models Eq. 5.3.

$$\left[PI = \left\{ \frac{R_t}{MAE_t + MRE_t + RMSE_t + PBIAS_t} * 0.6 \right\} + \left\{ \frac{R_T}{MAE_T + MRE_T + RMSE_T + PBIAS_T} * 0.4 \right\} \right]$$

(5.3)

where t is for testing and T is for training phase.

5.5   Case Study

The data of wave height, wind speed and water depth for the two locations were collected form the National Data Buoy Center. The most recent analysis dataset produced by the European Center for Medium-Range Weather Forecasts (ECMWF) [41]. The water quality data were taken from Fig. 5.2. The most recent reanalysis

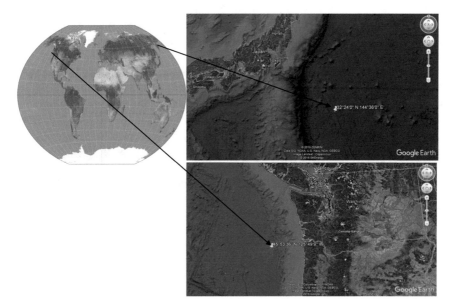

Fig. 5.2 Location of study sites

Table 5.1 Magnitude of the parameter with respect to the selected location

Parameter	USA (45° 53′ 36″N 125° 49′ 9″W)	Japan (32° 24′ 0″N 144° 36′ 0″E)
Wave height	5.9	7.5
Wind speed	7.8	8.6
Water depth	2289	5725
Wave period	7.6	8.2
Salinity	32	31

dataset produced by the European Centre for Medium-Range Weather Forecasts (ECMWF), it's used to study. The wave height (H_s) and wave period (T_e) are obtained from the spectral moment as shown in Eqs. (5.4) and (5.5)

$$\text{Significant of wave height } (\text{H}_s) = 4\sqrt{m_0} \tag{5.4}$$

$$\text{Energy period } (\text{T}_e) = \frac{m_{-1}}{m_0} \tag{5.5}$$

5.6 Result and Discussion

In the present study two criteria defined by efficiency and location was selected in the decision hierarchy system to determine the weight vector of feasibility location of wave energy potential. The pair wise comparison of each criteria was calculated by ANP method. The result of the MCDM method to estimated the weight vector of importance. The estimation of the parameters of the ANN model. The weight of the two successive iteration is shown in Table 5.2

The result of the Table 5.3 The score of the rank of the criteria based on the ANP method. The location and wave height found to be the most important criteria and efficiency and water depth were identified as the least important of criterion. We found the MCDM result, it can be observed that wave height (0.352) and water depth (0.110) the highest and lowest important with the location selection of wave energy potential.

The performance of the three models was #5ANPARC most reliable of other than existing models. The both models were trained by GMDH and MLRM. The models #5ANPNT and #5AMPMLRM were second and third most important developed in the study, the variable ranking along with output transformation and

Table 5.2 Score and rank of the criteria

Criteria	Score	Rank
Location	0.625	1
Efficiency	0.498	2

Table 5.3 Table showing weight vector of the alternatives of ANP method

Sl. No.	Alternative	Score
1	Wave height	0.352
2	Wind speed	0.246
3	Water depth	0.110
4	Salinity	0.292
5	Wave period	0.226

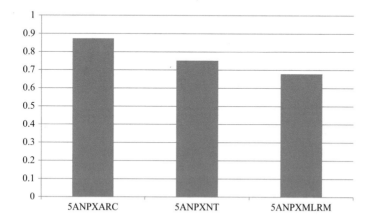

Fig. 5.3 Figure showing the results of the three models developed for prediction of wave energy potential

the another models trained with MLRM. All the models which had used five input variable were included.

If the method MCDM is changed the performance was selected models also changed in Fig. 5.3

The result from the case study analysis depicted Japan has the highest probability of wave energy potential is shown in Table 5.4.

The wave power potential per meter of wave crest of the two location was also calculated as recommended by [42] (Eq. 5.6)

$$P = \frac{\rho g^2}{64\pi} H_{mo}^2 T_e \approx \left(0.5 \frac{\text{KW}}{\text{m}^3 \text{ s}}\right) H_{mo}^2 T_e, \tag{5.6}$$

The USA and Japan, the power potential were found to be 3.63669 and 5.30515 MW/m³ s. The index value of USA and Japan was found to be equal to 0.27981 and 0.32157. The power potential and the index value were found to be the coherent with each other. It can be concluded the present index value accurate prediction of feasibility of wave energy potential.

Table 5.4 Table showing a index value of selected locations

Location name	Index value
Japan (32° 24′ 0″N 144° 36′ 0″E)	0.32157
USA (45° 53′ 36″N 125° 49′ 9″W	0.27981

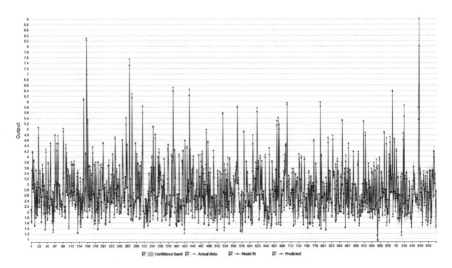

Fig. 5.4 Figure showing the distribution of residuals derived from observed and predicted output for #5ANPARC

The comparison of actual and predicted output during the training and testing phase in Fig. 5.4. The efficiency of the three models revealed that the developed #5ANPARC was the most consistent model among existing models. The most important sensitive model was trained by GMDH, the input and output was transformed the Arc tangent function and five variables were used as input.

5.6.1 Objective Equation

$$Y = -1.02872 + N917 * 0.36961 + N2 * 0.998769 \tag{5.7}$$

The five input and one output were developed to estimate the wave energy power potential in two location. The GMDH algorithm was used to train network where the value of weight vector of each input and number of hidden layer can be found in Eq. (5.7)

5.6.2 Multi Linear Regression Model

The objective equation of this model was developed by in Eq. (5.7)

$$Y = b0 + b1 * x + b2 * x + b3 * x + b4 * x + b5 * x \tag{5.8}$$

In the present investigation the MLRM trained model was found to have satisfactory performance metrics. A multiple linear regression was performed on the

Table 5.5 Table showing a equivalent performance index of three models

		PBIAS	MRE	MAE	RMSE	Correlation	Performance index (PI)	EPI
5ANPARC	Training	0.853274	0.022974	0.103947	0.162529	0.990649	7.0757205	0.873064476
	Testing	−6.66E−14	−1.81E−15	0.08509	0.108169	0.993118	9.261106	
5ANPBT	Training	−0.278305	−0.001029	0.13272	0.164832	0.98247	5.807574	0.750838576
	Testing	−3.14E−13	−8.774−15	0.11822	0.154007	0.98908	6.540466	
5ANPMLLRM	Training	0.514342	0.012521	0.08578	0.108043	0.984627	9.725654	0.678665
	Testing	−1.663E−13	−4.055E−15	0.07976	0.10088	0.99673	9.900674	

dataset to develop a relationship between the dependent and independent variables. The test of PI value of 0.678665 is less important of GMDH model algorithm. The value of correlation was calculated as 0.98462 indicating that the develop model was good fit for the dataset. The RMSE was calculated for both cases and was 0.108043 for the regression model compared to 0.162529 for the GMDH model. The GMDH model was better than the regression model in predicting in Table 5.5.

The test dataset was presented to both the regression model and the developed GMDH to validate the model. It should be noted that the GMDH model was never exposed to the test dataset during its training while the regression model was developed using the complete dataset, which included the test data points.

5.7 Conclusion

The present study attempts to predict the wave energy potential of different coastal regions. The study ANP MCDM and GMDH models were utilized to develop a three different models. The data representing various scenarios was generated and used to train the models. The Arc Tangent function was used in two cases to transfer the data of either input or output or both. Performance metrics like RMSE, MAE, PBIAS, R were used to find the equivalent performance of the models. The selected model was calculate the wave energy potential on study area. The accuracy of the model was found to be above 99.89%. The model output and the result from the power potential equation were compared and found to be coherent with each other, although magnitude of the results are well apart.

References

1. Chiari, L., & Zecca, A. (2011). Constraints of fossil fuels depletion on global warming projections. *Energy Policy, 39,* 5026–5034.
2. Hoel, M., & Kverndokk, S. (1996). Depletion of fossil fuels and the impacts of global warming. *Resource and Energy Economics, 18,* 115–136.
3. Nel, W. P., & Cooper, C. J. (2009). Implications of fossil fuel constraints on economic growth and global warming. *Energy Policy, 37,* 166–180.
4. Sebitosi, A. (2008). Energy efficiency, security of supply and the environment in South Africa: Moving beyond the strategy documents. *Energy, 33,* 1591–1596.
5. Ellabban, O., Abu-Rub, H., & Blaabjerg, F. (2014). Renewable energy resources: Current status, future prospects and their enabling technology. *Renewable and Sustainable Energy Reviews, 39,* 748–764.
6. Jayed, M., Masjuki, H., Kalam, M., Mahlia, T., Husnawan, M., & Liaquat, A. (2011). Prospects of dedicated biodiesel engine vehicles in Malaysia and Indonesia. *Renewable and Sustainable Energy Reviews, 15,* 220–235.
7. Mahlia, T., Abdulmuin, M., Alamsyah, T., & Mukhlishien, D. (2001). An alternative energy source from palm wastes industry for Malaysia and Indonesia. *Energy Conversion and Management, 42,* 2109–2118.

8. Ong, H., Mahlia, T., & Masjuki, H. (2012). A review on energy pattern and policy for transportation sector in Malaysia. *Renewable and Sustainable Energy Reviews, 16,* 532–542.
9. Ong, H., Mahlia, T., Masjuki, H., & Norhasyima, R. (2011). Comparison of palm oil, Jatropha curcas and Calophyllum inophyllum for biodiesel: A review. *Renewable and Sustainable Energy Reviews, 15,* 3501–3515.
10. Silitonga, A., Atabani, A., Mahlia, T., Masjuki, H., Badruddin, I. A., & Mekhilef, S. (2011). A review on prospect of Jatropha curcas for biodiesel in Indonesia. *Renewable and Sustainable Energy Reviews, 15,* 3733–3756.
11. Ayat, B. (2013). Wave power atlas of Eastern Mediterranean and Aegean seas. Energy 54:251–62.
12. Saket, A., & Etemad-Shahidi, A. (2012). Wave energy potential along the northern coasts of the Gulf of Oman, Iran. *Renewable Energy, 40,* 90–97.
13. Aoun, N. S., Harajli, H. A., & Queffeulou, P. (2013). Preliminary appraisal of wave power prospects in Lebanon. *Renewable Energy, 53,* 165–173.
14. Thorpe, T. W. (2000). The wave energy programme in the UK and the European wave energy Network. In *Fourth European wave energy Conference, Aalborg (Denmark).* Oxfordshire, UK: AEA Technology; October 2000.
15. Nielsen, K., Meyer, N. I. (1998). The Danish wave energy programme. In *3rd European Wave Energy Conference,* Patras, Greece.
16. Mollison, D., & Pontes, M. T. (2009). Assessing the Portuguese wave-power resource. *Energy, 17*(3), 255–268.
17. ESBI International. (2005). *Accessible wave energy resource atlas.* Report 4D404A-R2 for Marine Institute and Sustainable Energy Ireland, 2005.
18. Clément, A., McCullen, P., Falcao, A., Fiorentino, A., Gardner, F., Hammarlund, K., et al. (2002). Wave energy in Europe: Current status and perspectives. *Renewable and Sustainable Energy Reviews, 6*(5), 405–431.
19. Energy Portugal: Riding the wave of the future. Inter Press News Agency, 9/27/06. http://ipsnews.net/news.asp?idnews=34898. Accessed 10/19/06.
20. Bluhm-Lenee, P., et al. (2011). Characterizing the wave energy resource of the US Pacific Northwest. *Journal of Renewable Energy, 36,* 2106–2119.
21. Iglesias, G., et al. (2010). Wave energy resource in the Estaca de Bares area (Spain). *Journal of Renewable Energy, 35,* 1574–1584.
22. Dalton, G. J., et al. (2010). Case study feasibility analysis of the Pelamis wave energy convertor in Ireland, Portugal and North America. *Journal of Renewable Energy, 35,* 443–455.
23. Henfridsson, U., et al. (2007). Wave energy potential in the Baltic Sea and the Danish part of the North Sea, with reflections on the Skagerrak. *Journal of Renewable Energy, 32,* 2069–2084.
24. Iglesias, G., et al. (2011). Wave resource in El Hierrodan Island towards energy self-sufficiency. *Journal of Renewable Energy, 36,* 689–698.
25. Saaty, T. L. (1996). *Decision making with dependence and feedback: the analytic network process.* Pittsburgh, PA: RWS Publication.
26. Ahammeda, F., & Azeem, A. (2013). Selection of the most appropriate package of solar home system using analytic hierarchy process model in rural areas of Bangladesh. *Renewable Energy, 55,* 6–11.
27. Cheng, E., & Li, H. (2007). Application of ANP in process models: an example of strategic partnering. *Building and Environment, 42*(1), 278–287.
28. Saaty, T. L. (1999). Fundamentals of the analytic network process. In: *The International Symposium on the Analytic Hierarchy Process,* Kobe, Japan.
29. Willmott, C. J., & Matsuura, K. (2005). Advantages of the mean absolute error (MAE) over the root mean square error (RMSE) in assessing average model performance. *Climate Research, 30*(1), 79.
30. Saaty, T. L. (1990). How to make a decision: The analytic hierarchy process. *European Journal of Operational Research, 48,* 9–26.

31. Saaty, T. L. (2004). Fundamentals of the analytic network process—multiple networks with benefits, costs, opportunities and risks. *Journal of Systems Science and Systems Engineering, 13,* 348–379.
32. Hwang, H. S. (2006). Fuzzy GMDH-type neural network model and its application to forecasting of mobile communication. *Computers & Industrial Engineering, 50*(4), 450–457.
33. Mueller, J. A., Lemke, F. (2000). *Self-organizing data mining*, Libri, Berlin, 2000.
34. Ivakhnenko, A. (1971). Polynomial theory of complex systems. *IEEE Transactions on Systems, Man and Cybernetics, 4,* 364–378.
35. Kalantary, F., Ardalan, H., & Nariman-Zadeh, N. (2009). An investigation on the S_u–N_{SPT} correlation using GMDH type neural networks and genetic algorithms. *Engineering Geology, 109*(1), 144–155.
36. Despotovic, M., et al. (2016). Evaluation of empirical models for predicting monthly mean horizontal diffuse solar radiation. *Renewable and Sustainable Energy Reviews, 56,* 246–260.
37. Gray, J., et al. (2016). Thermodynamics of gas turbine cycles with analytic derivatives in OpenMDAO. In *57th AIAA/ASCE/AHS/ASC Structures, Structural Dynamics, and Materials Conference.*
38. Pascual-González, J., et al. (2016). Statistical analysis of the EcoInvent database to uncover relationships between life cycle impact assessment metrics. *Journal of Cleaner Production, 112,* 359–368.
39. Noori, N., & Kalin, L. (2016). Coupling SWAT and ANN models for enhanced daily streamflow prediction. *Journal of Hydrology, 533,* 141–151.
40. Gupta, H. V., Sorooshian, S., & Yapo, P. O. (1999). Status of automatic calibration for hydrologic models: Comparison with multilevel expert calibration. *Journal of Hydrologic Engineering, 4*(2), 135–143.
41. de Antonio, F. O. (2010). Wave energy utilization: A review of the technologies. *Renewable and Sustainable Energy Reviews, 14*(3), 899–918.
42. Ghosh, S., Chakraborty, T., Saha, S., Majumder, M., & Pal, M. (2016). Development of the location suitability index for wave energy production by ANN and MCDM techniques. *Renewable and Sustainable Energy Reviews, 59,* 1017–1028.

Chapter 6
A MCDM-NBO Approach for Selection of Installation Location for Wave Energy Power Plants

Tilottama Chakraborty

Abstract Wave energy is deemed to be one of the major options to substitute conventional energy sources. But due to the irregularity in wave patterns, technical shortcoming of converters, hostility from local populations and many other factors the cost of installation for wave power plant is expensive compared to other forms of renewable energy of the obstacles seems to be location dependent. That is why; selection of an ideal location for wave energy production can ensure optimal conversion of wave into electrical energy. The present study first tries to identify the most important parameters for location selection by MCDM methods. Then it attempts to identify the single most significant parameter for which production potential can be maximum. In this regard this study utilized three population based nature oriented optimization algorithms. The results from the study will reduce the search space of finding ideal location for installation of wave energy power plants. After applying the method in coastal region of Europe the results encouraged further applications of the method.

Keywords Wave energy · Nature based optimization · MCDM

6.1 Introduction

Due to the world's rising energy consumption and the limited existence of fossil energy sources, scarcity of energy is observed in many parts of the world. Wave energy is a renewable energy source with high power density, relatively high utilization factor, low visual impact and low impact on the environment compared to other non-conventional energy sources [1]. Wave energy could play a major part in the world's efforts to combat climate change, potentially displacing 1–2 billion tons of CO_2 per annum from conventional fossil fuel generating sources. Estimates

T. Chakraborty (✉)
Hydro-Informatics Engineering, National Institute of Technology Agartala,
Agartala 799046, Tripura (W), India
e-mail: tilottama86@gmail.com

© Springer Nature Singapore Pte Ltd. 2018
M. Majumder (ed.), *Application of Geographical Information Systems and Soft Computation Techniques in Water and Water Based Renewable Energy Problems*, Water Resources Development and Management, https://doi.org/10.1007/978-981-10-6205-6_6

121

consider that a potential of 10% of the global energy demand can be satisfied by generation from wave energy [2].

The total theoretical wave energy potential is estimated to be 32,000 TWh/yr (115 EJ/yr) [3], approximately twice the global electricity supply in 2008 (16,800 TWh/yr or 54 EJ/yr). According to the 2012 report World Energy Resources, the Western and Northern European, North, Central and South American, Asian, African and Australian coastal belts have potentials of 2800, 4000, 1500, 4600, 4600, 6200, 3500 and 5600 TWh/yr, respectively.

The global wave power potential is estimated to be in the order of 1–10 TW, which is in the same order of magnitude as the world consumption of electrical energy [4]. The total wave power found in Sweden is in the range of 1 GW, whereas in UK it is 120 GW [2]. Denmark, Portugal and Ireland have potentials of 3.4, 10, 21 GW, respectively [2]. Wave power generation is also found to be more reliable than other renewable resources, such as solar and wind energy, because its density $(2–3 \text{ kW/m}^2)$ is greater than the density of wind and solar (wind $0.4–0.6 \text{ kW/m}^2$; solar $0.1–0.2 \text{ kW/m}^2$).

Coastal areas all over the world are typically densely populated, and a large proportion of the coastline is already committed to other uses, such as commercial ports, fisheries and aquaculture, and leisure and sport activities.

Although there is a large potential to supply demand with wave energy, due to the irregularity in wave patterns, survivability, complexity in energy conversion due to diffraction and reflection, and so on, the cost of electricity generation via wave energy generation is still expensive [5] and has a maximum achieved efficiency of 90% [6]. Disturbance of floral and faunal life, navigability of oceans, and visual and noise pollution caused by the floating converters are some of the other obstacles that reduce the feasibility of wave power plants in many coastal regions even if there is sufficient power potential [7–9].

Most of the obstacles for which the conversion of wave energy becomes non-optimal depend on location. Again, as the data of all of the factors that are useful for identifying the ideal location for wave energy power plants with optimal utilization potential are sparse, the procedure for identifying locations that are most suitable for wave energy production is unreliable, difficulty and complex. Even if location selection factors are known, the regular collection of data for those factors is rather problematic and also extremely expensive. That is why the determination of the priority values of the related factors for which the conversion potential will be maximized is an important objective for identifying the most important parameter in the selection of locations for wave energy power plants. This will reduce the search space and search time for detection of an ideal location for wave energy production.

In this regard, the present investigation proposes a new method to determine the priority values of related factors for the identification of an ideal location at which the conversion efficiency of a wave energy converter (WEC) will be maximized. The methodology uses various MCDM methods to determine the domain for searching for the ideal priority value of the parameters. Then, optimization techniques are utilized to maximize the wave energy potential by varying the priority values and magnitude of the parameters for a set of considered locations. The new approach will reduce the data dependency and requirement of considering many factors for the selection of ideal locations for the installation of wave energy converters. The time and complexity involved in the determination of optimal locations for the said objective will also be significantly reduced. The applicability of the method was verified by identifying optimal locations for a set of study areas where the power potential is already known.

6.2 Wave Energy Potential in India

The implementation of wave energy is still in its pre-commercial phase [10], and much work is ongoing for the evaluation of wave energy potential; the detailed characterization of the wave energy potential is very important for all renewable energy resources.

India is estimated to have a potential of 40–60 GW of wave energy. The wave energy potential is estimated to be 5–15 MW per meter of coastline. India experimented with a 150 kW wave energy system at Thiruvananthapuram (Kerala) in 1983. The average wave potential along the Indian coast is roughly 5–10 kW/m. India has a coastline of approximately 7500 km. Even 10% utilization would mean a resource of 3750–7500 MW [DOD's; Annual Report (1990–91)].

6.3 Methods Adopted

Various methods were adopted to achieve the study objective. The Maximin, Minimax and Average Ranking methods were adopted to rank the alternatives based on the citation frequency, cost and efficiency potential of the selected parameters.

MCDM methods, such as AHP and ANP along with the Survival Function, were used to identify the domain of priority values for the selected parameters. The max and min magnitude of weights from the different weights proposed by the MCDM methods were used as the upper and lower limit of the search space for the priority values.

Again, optimization techniques, such as GA was used to maximize the objective function. The optimal point at which the function becomes the maximum yielded the ideal priority values for which the conversion potential will be maximized and for which the parameter that is most influential in maximizing the potential can be identified. In the next section, a brief description of the methods is given.

6.3.1 Ranking Method

6.3.1.1 Maximin

The maximin method looks for the maximum of the minimum value [11]. The minimum of the maximum value obtains the highest rank, whereas the minimum of the minimum value is taken as the lowest rank of all of the data available. The maximum approach is used to evaluate the relative efficiency of decision-making units (DMUs) with respect to multiple outputs and a single exact input with common weights.

6.3.1.2 Minimax

Minimax is a decision rule used in decision theory for minimizing the possible loss for a worst case (maximum loss) scenario. This method looks for the minimum of the maximum and ranks it the highest [11]. In taking the important decision-making problems in many manufacturing sectors, the Minimax method is used. Minimax is used in optimal Inference from Partial Rankings.

6.3.1.3 Average Ranking Method

This is a simple ranking method via Friedman's M statistics. The average ranking method was used to develop the Fuzzy Weighted Average for Ranking Alternatives. In the optimization of energy consumption in Iran, this method was used to rank the criteria on which the objective is dependent.

6.3.2 Multi-criteria Decision Making Method

6.3.2.1 Analytical Hierarchy Process (AHP)

The Analytical Hierarchy Process (AHP) is a multicriteria decision making method (MCDM) introduced by Saaty [12] based on the relative priorities assigned to each criterion's role in achieving the objective. Whenever a goal for a decision can be clearly stated, a set of relevant criteria can be determined and a set of alternatives can be described using these criteria; AHP is an appropriate tool for these problems. In this method, the problem is breaking down a problem into smaller and smaller consistent parts. The best alternative is usually selected by making comparisons between alternatives with respect to each attribute.

The ability of AHP to process flexibility, such as criteria selection, technology selection and criteria weightage, allows for its use by decision makers [13]. The main advantage of this method is that it can handle a complex problem by preparing a hierarchy of choices and explaining the reasons for such choices through decomposition and synthesis. The advantages of AHP over other multi-criteria methods are its flexibility, intuitive appeal to decision makers and ability to check inconsistencies. Additionally, the AHP method has the distinct advantage that it decomposes a decision problem into its constituent parts and builds hierarchies of criteria. The AHP method supports group decision which is made by calculating the geometricmean of the individual pairwise comparisons.

The AHP method can be considered to be a complete aggregation method of the additive type. The problem with such aggregation is that compensation between good scores on some criteria and bad scores on other criteria can occur. Detailed, and often important, information can be lost by such aggregation. In AHP, the decision problem is decomposed into a number of sub-criteria, within which and between which a substantial number of pairwise comparisons need to be completed. This approach has the disadvantage that the number of pairwise comparisons to be made may become very large and thus become a lengthy task.

Luthra et al. [14] states that the Analytical Hierarchy Process (AHP) technique can be utilized in terms of the ranking of barriers to adopt renewable sustainable technologies in the Indian context. Kolios et al. [15] states the use of analytical multi-criterion analysis for the prioritization of risks for the development of tidal energy projects. Abdullah and Najib [16] shows a new concept of IF-AHP, which is applied to establish a preference in the sustainable energy planning decision-making problem. AHP is also applicable in the selection of renewable energy sources for the sustainable development of electricity generation systems [17]. For the selection of a methodology of hydrogen energy storage, a combined Fuzzy-AHP was used. Chen and Chen [18] states a conceptual model for the strategic planning of energy sources and its critical decision-making factors, and then introduces AHP for the analysis of benefits, opportunities, costs and risks to select the suitable planning and policy for energy sources. Shen et al. (2011) states the use of AHP in a proposed method of a portfolio of renewable energy sources to achieve energy, environmental, and economic policy goals for Taiwan. Kaya et al. (2010) shows multi-criteria renewable energy planning using an integrated fuzzy VIKOR and AHP methodology. Kaya et al. (2010) used AHP for multi-criteria renewable energy planning.

6.3.2.2   Analytical Network Process (ANP)

The ANP is the general form of the analytic hierarchy process (AHP) [12] and has been used in multi-criteria decision making (MCDM) to calculate priorities. This method consists of two parts. The first one is a control hierarchy of criteria and sub-criteria that control the feedback network. The second part consists of the networks of influence that contain the factors of the problem and the logical groupings of these factors into clusters.

For studying more complex decision problems, ANP allows for interactions and influences among the various components of the decision problem, which makes it a better choice [19]. In ANP, not only the importance of the criteria determines the importance of the alternatives, as in a hierarchy, but the importance of the alternatives impact on the importance of the criteria is also included [20]. It is more time effective than AHP [21].

Kabak and Dagdeviren [22] uses ANP for the prioritization of renewable energy sources for Turkey. Hasanzadeha et al. [23] states the application of ANP in the selection of an environmental site for an oil jetty. ANP can also be applied to find a suitable location of a particular solar-thermal power plant project [24]. The ANP technique is used to assess the Green Energy alternatives [20]. Atmacaa et al. (2012) determines the suitability of existing power plants in Turkey and the plants that are being considered for establishment in the near future using ANP. Liu et al. (2010) shows the application of ANP in environmental impact evaluation studies.

6.3.2.3 Fuzzy Decision Making

L.A. Zadeh set the basis of fuzzy set theory as a process to address the haziness of practical systems in 1965 [25]. Fuzzy Logic (FL) is a logic that allows intermediate values to be defined between conventional evaluations, such as true/false, yes/no, high/low, and so on. The main benefits of fuzzy logic include its simplicity and its flexibility. Fuzzy logic can handle problems with imprecise and incomplete data, and it can model nonlinear functions of arbitrary complexity. It describes any system in terms of a combination of linguistics and numbers. The main disadvantage of FDM is that, for a crisp input and output system, using fuzzy becomes a different way of performing interpolation. For a more precise system, this method can take more time than any other conventional system. Elgammal et al. (2014) show the application of Fuzzy Logic in a closed-loop vector control structure based on adaptive Fuzzy Logic Sliding Mode Controller (FL-SMC) for a grid-connected Wave Energy Conversion System (WECS). Kim et al. (2014) used the Fuzzy model to forecast the offshore bar-shape profiles under high waves. In the study of predictions of sea level with different forecast horizons, the fuzzy model was used [26]. Ozger (2010) shows the application of the fuzzy model in the forecasting of significant wave height in his study.

6.3.2.4 AHP-Decision Making Trial and Evaluation Laboratory (DEMATEL)

The AHP DEMATEL method, originated from the Geneva Research Centre of the Battelle Memorial Institute, is especially pragmatic to visualize the structure of complicated causal relationships. DEMATEL is a comprehensive method for building and analyzing a structural model involving causal relationships between complex factors. It can clearly expose the cause effect relationship of criteria when measuring a problem. It shows a basic concept of contextual relation among the elements of the system (which is not a part of this study because of its integrated methodology in which the numeral represents the strength of the influence). AHP was used to rate the alternatives, with the criteria weighted per importance by DEMATEL. DEMATEL is a good technique for evaluating problems; the relationships of systems are generally given by crisp values when establishing a structural model. The major disadvantages of DEMATEL is that the comparison scale doesn't show the ambiguities of human assessments, and due to the subjectivity of expert judgment and the complexity of things, it is very difficult for an individual expert to provide accurate judgmental information. Chen et al. [27] established a decision model for improving the performance of solar farms using DEMATEL. It can also be used to select the best project for the environment planning of coastal wetlands regions and for the measurement and evaluation of environmental watershed plans (Chen et al. [28]. Chang and Cheng [29] evaluates the risk of failure using DEMATEL.

6.3.2.5   Survival Function

The survival function, also known as a reliability function, is a property of any random variable that maps a set of events, usually associated with the mortality or failure of some system, to time. It captures the probability that the system will survive beyond a specified time. The output from the best three distribution functions that can represent the data pattern accurately were used to estimate the survival of the parameter importance with respect to the Minimax, Maximin and Average Ranking Methods. This is the first time where the survival function is used in decision making objectives.

6.3.2.6   Optimization Technique (OT)

OT is a group of methods that are applied to find the maximum value of the objective function within the domain specified by the highest and lowest priority values and the highest and lowest magnitude of the parameters for the set of considered locations. Algorithms such as GA was used to maximize the proposed objective functions.

Genetic Algorithms (GA)

GA is an example of a nature-based OT first proposed by Holland and further developed by Goldberg. GAs simulate the survival of the fittest among individuals over consecutive generations to solve a problem. Each generation consists of a population of character strings that are analogous to the chromosomes that we see in our DNA. Each individual represents a point in a search space and a possible solution. The individuals in the population are then made to go through a process of evolution.

A genetic algorithm is very easy to understand. It gives multiple solutions of a given problem. As the execution of this method is not dependent on the error surface, it can solve multi-dimensional, non-differential, non-continuous, and even non-parametrical problems.

Variant problems cannot be solved by means of genetic algorithms, and there is no absolute assurance of finding a global optimum with this method. It is unreasonable to use genetic algorithms for online controls in real systems without testing them first on a simulation model.

Genetic algorithms can be used to develop a robust, systematic method of optimizing the collector shape to improve energy extraction [30]. Saravanan and Balakrishnan [31] shows the application of GA in the design of a renewable energy-based shunt active filter with a multilevel inverter analyzed the performance of GA for solving the problem of minimum cost expansion of power transmission networks.

Particle Swarm Optimization (PSO)

The particle swarm optimization (PSO) algorithm is based on the evolutionary computation technique. PSO optimizes an objective function by conducting a population based search. The population consists of potential solutions, called particles, similar to birds in a flock. The particles are randomly initialized and then freely fly across the multi-dimensional search space. While flying, every particle updates its velocity and position based on its own best experience and that of the entire population. The updating policy will cause the particle swarm to move towards regions with higher objective values. Eventually, all of the particles will gather around the point with the highest objective value. PSO attempts to simulate social behavior, which differs from the natural selection schemes of genetic algorithms.

The main advantage of PSO is that it is easy to implement and the number of adjusted parameters is much smaller. PSO can be applied in both scientific research and engineering uses, as calculation in PSO is very simple. Compared with other calculations, PSO has greater optimization ability and can be completed easily. If the regulation of its speed and direction are less exact, this method easily suffers from partial optimization. It cannot be used for problems on non-coordinate systems.

Elgammal et al. (2014) gave information on his attempts to use PSO to develop a closed-loop vector control structure based on an adaptive Fuzzy Logic Sliding Mode Controller (FL-SMC) for a grid-connected Wave Energy Conversion System (WECS). Swarm intelligence and gravitational search algorithm can also be used for optimization of synthesis gas production (Ganesan and Elamvazuthi [32].

Jia Yuge et al. (2012) adopted an improved particle swarm optimization, i.e., an adaptive particle swarm optimization, for wave impedance inversion. Elgammal (2014) shows the optimal design of a PID controller for a doubly fed induction generator-based wave energy conversion system and also presents a standalone hybrid diesel-wave energy conversion system using PSO.

Differential Evolution Algorithm (DE)

The DE algorithm is a population-based algorithm, like genetic algorithms, that uses similar operators: crossover, mutation and selection. The main difference in constructing better solutions is that genetic algorithms rely on crossover while DE relies on mutation. This main operation is based on the differences of randomly sampled pairs of solutions in the population. Differential Evolution (DE) is an evolutionary algorithm (EA) that was developed to handle optimization problems over continuous domains.

The advantages of DE are mainly fast convergence and the use of a small number of control parameters. The main advantages of differential evolution are that it often displays better results than other evolutionary algorithms. It is easily applied to real valued problems.

The main disadvantage of this method is that the convergence is unstable.

Amjad et al. [33] shows the application of DE for a multilevel voltage source inverter. Differential Evolution (DE) can also be applied in renewable energy conversion systems (Amjad and Salam [34]). Jiang used an improved adaptive differential evolution algorithm for parameter estimation of solar cells and modules. Hasan Doagou-Mojarrad [35] solve the problem of the Multi-objective optimal placement and sizing of DG (distributed generation) units in the distribution network applying DE. Frank et al. (2012) applies the DE algorithm to identify the optimal power flow method proposed a methodology that employs the differential evolution (DE) algorithm to find the optimal settings of RPD control variables.

6.4 Proposed Methodology

The proposed model for the identification of the most suitable location for wave energy production is composed of the following steps:

Step-1: Application of MCDM to identify the range of priority values by which the importance of the parameters can be represented.

Step-2: The objective function to represent the wave energy potential was determined.

Step-3: The nature-based optimization was used to maximize the function for different locations and different weights of importance for the parameters.

Step-4: Once the optimal value of the objective function was identified, the parameter with the highest weights was considered as the highest priority and as the most important parameter for determination of the ideal location for installation of the wave energy power plant.

The method was applied to find the optimal location among a set of three locations in the European coastal belt. The detailed methodology and the results are discussed in Sect. 6.4.1.

6.4.1 Detailed Methodology

In the present method, the most important parameter for the identification of wave energy was identified by the application of different optimization and MCDM techniques. The objective function was proposed in such a manner that, at a certain priority value and magnitude of the parameters, the utilization potential will be maximized. A range of priority values for each parameter is determined by the application of different MCDM methods. The domain for the magnitude of the parameters was determined from the data collected from the considered locations.

In the MCDM methods, three sub-criteria were used to represent the equivalent impact of three different criteria on the parameters.

The criteria considered are critical frequency in related studies and the cost and efficiency potential of the parameters. The cumulative impact of the three criteria was depicted by the results from the Minimax, Maximin and Average Ranking Method. Then, MCDM methods, such as AHP, ANP, FDM and DEMATEL were used to find the maximum and minimum relative weights or priority values of the parameters. In the next section, the method by which values of the criteria and sub-criteria were determined is explained with examples.

6.4.1.1 Citation Frequency in the Related Literature

Various studies were surveyed to seek out citations of all of the parameters in connected studies. If the range of studies that mentioned the parameter is c and the total number of literatures surveyed is C, then the score, the Survey of the literature (SL), is calculated by Eq. 6.1

$$SL = (c/C) \tag{6.1}$$

In this study, 30 total studies on wave energy, including wave energy potential, converter selection and converter design, were reviewed. Within the 30 works, each considered significant wave height as the most important parameter. Therefore, the SL for significant wave height is (30/30) = 1 (100%) according to Eq. 6.1.

The SL of the other six parameters was also calculated in a similar manner.

6.4.1.2 Efficiency Potential

The commonly used equation for calculating the power potential, as proposed by Pontes et al. (1995) and Tucker and Pitt (2001), is given in Eq. 6.2.

$$P_w = \frac{\rho g^2}{64\pi} T_e H_s^2 \tag{6.2}$$

where

P_w Average wave power;
H_s^2 Significant wave height;
T_e Peak period;
ρ Density of water and;
g Accelaration due to gravity.

The parameters that are directly proportional to the Wave Power Potential were considered to be conducive, and the factors that are increasingly proportional to the power potential were taken as deductive for the conversion efficiency of wave power plants. The equation of the power potential (Eq. 6.2) was used in this regard to estimate the score of the alternatives with respect to the efficiency potential. The equation of power potential states that P_w (power production) is directly proportional with T_e (time to peak) and H_s^2 (significant wave height). Again, H_s is also dependent on wave amplitude, wind speed, duration and fetch. T_e is again a function of ocean depth. If all of the alternatives are ranked according to their proportionality with the power potential, then Table 6.1 can show the relative importance for each of the parameters with respect to the Efficiency Potential.

As H_s squared is directly proportional to P_w, the efficiency potential or location with a high magnitude wave height will have a higher level of conversion efficiency. The relative score was calculated by Eq. 6.3

$$RS = \left(\frac{R}{MaxR}\right)^{-1} \tag{6.3}$$

where $R = 1, 2, \ldots, 7; Max(R) = 7$.

Table 6.1 shows the relative score of each of the considered parameters.

6.4.1.3 Cost Potential

The cost potential of parameters depends on the proportionality of the parameter to the mooring cost or the cost required to hold the converters in place. The score of

Table 6.1 The relative importance for each of the parameters with respect to the efficiency potential

Alternative	Proportionality with power potential	Rank of importance (R)	Relative Score (RS = R/Max R)$^{-1}$	Final rank
Significant wave height (H$_s$)	$P \propto H_s^2$	2	3.50	2
Wave amplitude (a)	$P \propto 4\,H_s^2$	1	7.00	1
Peak period (T$_e$)	$P \propto T_e$	3	2.33	3
Wind duration (WD)	$H_s \propto WD$	4	1.75	4
Depth of the ocean (OD)	$T_e \propto OD$	7	1.00	7
Fetch (F)	$H_s \propto F$	4	1.75	4
Wind speed (WS)	$H_s \propto WS$	4	1.75	4

the parameters for the cost potential is calculated by Eq. 6.4. If ΔC is the difference in cost for two different locations and ΔH_s is the change in the wave height, then the cost potential of wave height can be represented by Eq. 6.4

$$C = \frac{\Delta C}{\Delta H} \tag{6.4}$$

Now, the higher the value of C, the lower the rank of the alternatives. That is why if the C of the Significant wave height is more than the C of the wind speed, then the significant wave height will have a lower rank than wind speed. The general equation for the estimation of the cost potential for the parameter is depicted in Eq. 6.5

$$c = \frac{\Delta C}{\Delta P} \tag{6.5}$$

where ΔP is the change in the magnitude of the parameter with respect to locations (Fig. 6.1).

6.4.1.4 Selection of Alternatives

Three sub-criteria were used to compare the weights of the importance for the parameter with respect to the results of the criteria considered. Minimax, Maximin and the average ranking method were used to rank the seven parameters based on the Survey of literature, Efficiency and Cost Potential. Table 6.2 shows the ranks of the parameters for the three sub-criteria.

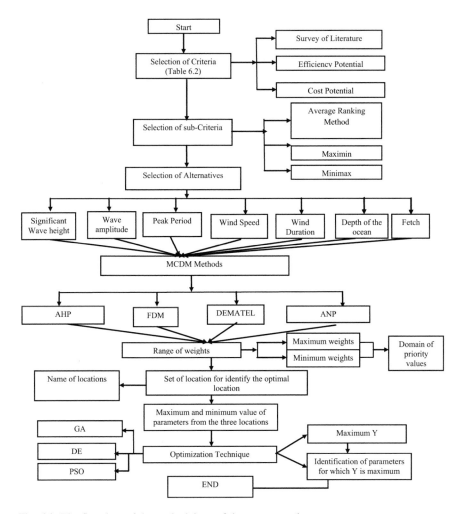

Fig. 6.1 The flowchart of the methodology of the present study

Table 6.2 The rank of each parameter with respect to the selected criteria

Sl. No.	Parameter	Rank		
		SL	Efficiency	Cost
1	Significant wave height	1	2	5
2	Wave amplitude	1	1	3
3	Peak period	1	3	3
4	Wind duration	1	4	2
5	Depth of the ocean	1	7	1
6	Fetch	1	4	7
7	Wind speed	1	4	6

6.4.1.5 Aggregation Method

The ranks of the alternatives were determined with the help of survey of the literature, efficiency and cost. However, the weights of importance or ranges of weights of the parameters were estimated with the help of MCDM and the survival function based on the rankings of the alternatives received with respect to the Maximin, Minimax and Average ranking methods. Table 6.3 shows the ranks of alternatives with respect to the Maximin, Minimax and the Average ranking methods. This rank shows the equivalent impact of the selected criteria on the importance ranks of the alternatives. AHP, ANP, FDM, DEMATEL and the survival function were used to determine the relative weights of the alternatives. The results of the methods are depicted by Table 6.4.

The maximum and minimum weights of each parameter are depicted in Table 6.4. According to the table, the wave amplitude and fetch are the most important and least important parameters, respectively.

6.4.1.6 Study Area

The applicability of the method was validated by applying the method on a set of three locations. Table 6.5 shows the magnitude of the parameters in these three locations. It should be noted that the highest and lowest wave amplitude and wave height among the selected locations are found in the Wave Hub test site and Port Kambla respectively. Table 6.5 also shows the power potential of the three locations. It can be observed that the power potential of the Wave Hub is the maximum, whereas that of Port Kembla is the minimum.

Biscay Marine Energy Platform (bimep), Spain: The Biscay Marine Energy Platform (bimep) test site is located in the sea off the coast of the village of Armintza in the municipal area of Lemoiz, some 30 km north of Bilbao in the Basque Country, Spain [36]

Port Kembla, Australia: Port Kembla, Australia is the site where the Oceanlinx OWC system was tested on the east side of the country. The Port Kembla Wave Energy Barge is located approximately 80 m offshore from Rockwall Road, Port Kembla [36].

Table 6.3 The rankings of the parameters with respect to the selected sub-criteria

Sl. No.	Name of the parameters	Weights		
		Average ranking method	Maximin	Minimax
1	Significant wave height	4	3	3
2	Wave amplitude	1	1	1
3	Peak period	2	2	2
4	Wind duration	2	4	4
5	Depth of the ocean	5	6	6
6	Fetch	7	6	6
7	Wind speed	6	4	4

Table 6.4 The relative weights or priority values of each parameter according to the different selected MCDM and PDF methods

Name of the parameters	Weight vector				Survival function			Priority value		
	AHP	ANP	FDM	DEMATEL	Best function 1	Best function 2	Best function 3	Average	Max	Min
Wave amplitude	0.387	0.865	0.176	0.328	0.362	0.343	0.349	0.401	0.865	0.176
Significant wave height	0.130	0.290	0.265	0.129	0.332	0.328	0.349	0.260	0.349	0.129
Wind speed	0.060	0.144	0.083	0.037	0.231	0.926	0.227	0.244	0.926	0.037
Peak period	0.205	0.452	0.136	0.195	0.219	0.275	0.064	0.221	0.452	0.064
Wind duration	0.118	0.226	0.134	0.179	0.242	0.265	0.240	0.201	0.265	0.118
Depth of the ocean	0.057	0.113	0.112	0.084	0.288	0.294	0.269	0.174	0.294	0.084
Fetch	0.043	0.093	0.095	0.048	0.216	0.220	0.207	0.132	0.220	0.043

Table 6.5 The magnitudes of the top five important parameters at the selected location

Parameters	Bimep, Spain (43.28°N, 2.51°W)	Port kembla (34.27°E, 15.54°S)	Wave hub (50.36° N, 5.67°W)
Significant wave height (m)	11.45	6.6	14.4
Wave amplitude (m)	5.725	3.3	7.2
Peak period (s)	15.4	10	14.1
Wind duration (h)	5	2	5
Wind velocity (m/s)	47	50	33.2
Power potential (kW) per meter of wave crest	88.12	33.00	101.52

Wave Hub: The Wave Hub test site is a project in the Southwest of England, located 16 km offshore near Cornwall St. Ives Bay [36].

6.4.1.7 Determination of the Objective Function for Optimization

Once the domain for the priority values and the range of the selected locations and their magnitudes for the selected parameters were identified, the next and final step was to identify the priority values and magnitudes of the parameters at which the potential will be maximized. Equation 6.6 depicts the objective functions developed to represent the utilization potential of wave energy.

The nature-based optimization techniques, such as GA was used to maximize the equation. The maximum and minimum priority values are used as the upper and lower limit of the search space for the priority values of the parameters. Similarly, the maximum and minimum values of the magnitudes in the selected locations were considered to be the search domain for varying the magnitudes of the parameters.

As all of the programming techniques are population-based, three different populations were selected to maximize the equation, and in total, the function is maximized by nine different variations of the three optimization algorithms.

$$
\begin{aligned}
F(x) = & ((W_{1max} - W_{1min}) \times Rand(0,1)) \times (V_{1min} + (V_{1max} - V_{1min}) \times Rand(0,1)) \\
& + (W_{2max} - W_{2min}) \times Rand(0,1)) \times (V_{2min} + (V_{2max} - V_{2min}) \times Rand(0,1)) \\
& + (W_{3max} - W_{3min}) \times Rand(0,1)) \times (V_{3min} + (V_{3max} - V_{3min}) \times Rand(0,1)) \\
& + (W_{4max} - W_{4min}) \times Rand(0,1)) \times (V_{4min} + (V_{4max} - V_{4min}) \times Rand(0,1)) \\
& + (W_{5max} - W_{5min}) \times Rand(0,1)) \times ((V_{5min} \times (V_{5max} - V_{5min}) \times Rand(0,1))
\end{aligned}
$$

$$(6.6)$$

where

V_1 Wave amplitude,
V_2 Significant wave height,
V_4 Wind speed, and

V_5 Wind duration, and
W_1 Weight values of the wave amplitude,
W_2 Weight values of the significant wave height,
W_3 Weight values of the wind speed,
W_4 Weight values of the peak period, and
W_5 Weight values of the wind duration.

Table 6.6 shows the maximum and minimum value of each of the parameters for the locations considered, and Table 6.7 shows the variables, constraints, population size, no of function evaluations and value of the maximum objective function according to the population size of the three different algorithms.

Table 6.7 shows that the maximum objective function value was found by the PSO technique at a population size of 30, and Table 6.8 shows the corresponding optimal values of each of the parameters and the weights. The maximum weight value was found to be 0.482, which is a parameter of wave amplitude that shows that the wave amplitude is the most important parameter for the wave energy production potential. If the wave amplitude is maximized in any location, then the potential of wave energy production can also be maximized.

The magnitude of the parameters for the optimal value of Eq. 6.6 depicts the ideal location where the conversion potential will be maximized. When the ideal value was compared with the three selected locations, it was found that Wave Hub,

Table 6.6 Minimum and maximum values of the parameters of the locations considered

Parameters	Maximum value	Minimum value
Significant wave height (m)	14.4	6.60
Wave amplitude (m)	7.20	3.30
Peak period (s)	15.4	10.0
Wind duration (h)	5.00	2.00
Wind velocity (m/s)	50.0	6.00

Table 6.7 Programming techniques used

Variables	Weights (W_n)	Parameter (V_n)
Constraints	Minimum and maximum values of weights predicted by 4 MCDM and survival function	Maximum and minimum magnitude of each of the parameter from the three locations
Programming techniques	GA	
Population size	25–200	
No. of the function evaluation	200,000	
Value of objective function (**Maximum** $F(x)$)	0.767 (for population size 100)	
Iteration No.	12	

Table 6.8 The magnitude and priority value of the parameters for which Eq. 6.5 becomes the maximum

Parameters	Optimal magnitude	Weights	Priority value
Wave amplitude	7.20	W_1	0.481
Significant wave height	15.41	W_2	0.249
Wind speed	35.00	W_3	0.072
Peak period	14.33	W_4	0.056
Wind duration	5.00	W_5	0.032

England, is nearest to the location where the converter potential will be maximized. The power potential of the location is also found to be maximized compared to the other two locations. The priority value of Wave amplitude and Significant Wave height is found to be the maximum and next to the maximum, respectively, whereas the weights for wind duration were found to be minimized. That concludes the importance of wave amplitude and Significant Wave height in the determination of location suitability for the optimal magnitude of the wave power potential.

The new method is novel in three aspects. The survival function was used for the first time to determine priority values for the parameters. Cost and Efficiency Potential was also used for the first time as criteria in MCDM applications.

6.5   Conclusion

The present study aimed to estimate the potential of the wave energy production of any given location. The different MCDM methods and Probability distribution function were applied, and the ensemble of all of the methods was used to determine the weights of the selected parameters with respect to their influence on the wave energy potential. The Optimization technique was cascaded with the weight value of each parameter and the range value of three locations to estimate the optimal value of selected parameters and the weight values, respectively. According to the results, wave amplitude was found to be the parameter of highest importance. The wave power equation was used to validate the most important parameter.

This research should be studied in the future with the three different location ranges as offshore, onshore and near shore. Then, the different optimal value for each of the different locations can be identified and used to estimate the maximum power output. Additionally, the other MCDM and OT methods can also be considered to identify the most important parameter for each of the locations. This study is mainly based on the wave spectrum, and the cost is considered only based on the mooring analysis. Other costs and other factors, such as environmental impact, power capacity and water quality, can also be included to identify the other important parameters.

References

1. Leijon, M., Bernhoff, H., Berg, M., & Agren, O. (2003). Economical considerations of renewable electric energy production especially development of wave energy. *Renewable Energy, 28*(8), 1201–1209.
2. Iglesias, G., Carballo, R. (2010). Wave energy resource in the Estaca de Bares area (Spain). *Renewable Energy, 35*(7).
3. Mork, G., Barstow, S., Pontes, M. T., & Kabuth, A. (2010). Assessing the global wave energy potential. In: 29th International Conference on Ocean Proceedings of OMAE (ASME), Offshore Mechanics and Arctic Engineering, Shanghai, China, 2010.
4. Mondial, C., Leneregei, D. E. (2013). World energy resources survey for sustainable energy.
5. de Falcao, A. F. O. (2010). Wave energy utilization: A review of the technologies. *Renewable and Sustainable Energy Reviews, 14*(3).
6. Drew, B., Plummer, A. R., & Sahinkaya, M. N. (2009). A review of wave energy converter technology. *Journal of Power and Energy.*
7. Boehlert, G. W. (2008). Ecological effects of wave energy development in the Pacific Northwest. NOAA Technical Memorandum NMFS-F/SPO-92.
8. Parnell, K. E., Kofoed-Hansen, H. (2001). Wakes from large high-speed ferries in confined coastal waters: Management approaches with examples from New Zealand and Denmark. *Journal of Coastal Management, 29*(3).
9. Pelc, R., & Fujita, R. M. (2002). Renewable Energy from the ocean. *Journal of Marine Policy, 26*(6), 471–479.
10. Banosa, R., Manzano-Agugliarob, F., Montoyab, F. G., Gila, C., Alcaydeb, A., Gomezc, J. (2011). Optimization methods applied to renewable and sustainable energy: A review. *Renewable and Sustainable Energy Reviews, 15*(4).
11. Berman, O., & Wang, J. (2007). The 1-minimax and 1-maximin problems with demand weights of general probability distributions. *Networks, 50*(2), 127–135.
12. Saaty, T. L. (1980). *The analytic hierarchy process.* NY: McGraw-Hill Book Co.
13. Jain, R., & Rao, B. (2015). Application of AHP tool choosing a medical research area. In *6th European Conference of the International Federation for Medical and Biological Engineering IFMBE Proceedings* (Vol. 45, pp. 1004–1007).
14. Luthraa, Sunil, Kumarb, Sanjay, Gargc, Dixit, & Haleemd, Abid. (2015). Barriers to renewable/sustainable energy technologies adoption: Indian perspective. *Renewable and Sustainable Energy Reviews, 41,* 762–776.
15. Kolios, A., Reada, G., & Ioannou, A. (2014). Application of multi-criteria decision making to risk prioritisation in tidal energy developments. *International Journal of Sustainable Energy, 27.*
16. Abdullah, L., & Najib, L. (2016). Sustainable energy planning decision using the intuitionist fuzzy analytic hierarchy process: Choosing energy technology in Malaysia. *International Journal of Sustainable Energy, 4,* 360–377.
17. Zhu, Q., Lujiaa, F., Mayyasa, A., Omara, M. A., Al-Hammadib, Y., Al Salehb, S. (2014). Production energy optimization using low dynamic programming, a decision support tool for sustainable manufacturing. *Journal of Cleaner Production.*
18. Chen, H. H., & Chen, S. (2013). The conceptual model for the strategic planning of energy sources.
19. Tjadera, Y., Maya, J. H., Shanga, J., Vargasa, L. G., Gaob, N. (2014). Firm-level outsourcing decision making: A balanced scorecard-based analytic network process model. *International Journal of Production Economics, 147*(Part C), 614–623.
20. Oztayşi, B., Ugurlu, S., & Kahraman, C. (2013). Assessment of green energy alternatives using fuzzy ANP. In *Assessment and simulation tools for sustainable energy systems green energy and technology* (Vol. 129, pp. 55–77).

21. Hernandez, C. T., Castro, R. C., Marins, F. A. S., Duran, J. A. R. (2012). Using the analytic network process to evaluate the relation between reverse logistics and corporate performance in Brazilian companies. *Revista Investigation Operational, 33*(1).
22. Kabak, M., & Dagdeviren, M. (2014). Prioritization of renewable energy sources for Turkey by using a hybrid MCDM methodology. *Energy Conversion and Management, 79,* 25–33.
23. Hasanzadeha, M., Boushehr, I., Danehkarb, A. (2014). Environmental site selection for oil jetty using the analytical network process method case study. *Ocean Engineering, 77,* 55–60.
24. Aragonés-Beltrán, P., Chaparro-González, F., Pastor-Ferrando, J. P., & Pla-Rubio, A. (2014). An AHP (analytic hierarchy process)/ANP (analytic network process)-based multi-criteria decision approach for the selection of solar-thermal power plant investment projects. *Energy, 66,* 222–238.
25. Zadeh, L. A. (1965). *Fuzzy sets.* In *Information and Control, 8.*
26. Karimia, Sepideh, Kisib, Ozgur, Shiria, Jalal, & Makarynskyyc, Oleg. (2013). Neuro-fuzzy and neural network techniques for forecasting sea level in Darwin Harbor. *Australia, Computers & Geosciences, 52,* 50–59.
27. Chen, C.-R., Huang, C.-C., & Tsuei, H.-J. (2014). Research article: A hybrid MCDM model for improving GIS-based solar farms site selection. *International Journal of Photo energy, 2014,* Article ID 925370.
28. Chen, V. Y. C., Tzeng, G.-H. (2010). The best project selection for the environment planning of coastal wetlands region based on a hybrid MCDM model.
29. Chang, K.-H., & Cheng, C.-H. (2011). Evaluating the risk of failure using the fuzzy OWA and DEMATEL method. *Journal of Intelligent Manufacturing, 22*(2), 113–129.
30. McCabe, A. P. (2013). Constrained optimization of the shape of a wave energy collector by genetic algorithm. *Renewable Energy, 51,* 274–284.
31. Saravanan, P., & Balakrishnan, P. (2013). Design of Renewable energy based shunt active filter with multilevel inverter using genetic algorithm. *International Journal of Engineering.*
32. Ganesan, T., Elamvazuthi, I., Shaari, K. Z. K., Vasant, P. (2013). Swarm intelligence and gravitational search algorithm for multi-objective optimization of synthesis gas production. *Applied Energy, 103,* 368–374.
33. Amjad, A. M., Salam, Z., & Saif, A. M. A. (2015). Application of differential evolution for cascaded multilevel VSI with harmonics elimination PWM switching. *International Journal of Electrical Power & Energy Systems, 64,* 447–456.
34. Amjad, A. M., & Salam, Z. (2014). A review of soft computing methods for harmonics elimination PWM for inverters in renewable energy conversion systems. *Renewable and Sustainable Energy Reviews, 33,* 141–153.
35. Doagou-Mojarrad, H., Gharehpetian, G. B., Rastegar, H., Olamaei, J. (2013). Optimal placement and sizing of DG (distributed generation) units in distribution networks by novel hybrid evolutionary algorithm. *Applied Energy, 54,* 129–138.
36. Ramboll. (2010). ANNEX II Task 1.1 Generic and Site-related Wave Energy Data.

Chapter 7
Cost Optimization of High Head Run of River Small Hydropower Projects

Sachin Mishra, S.K. Singal and D.K. Khatod

Abstract Small hydropower (SHP) is one of the most reliable and environment friendly source of electrical power among different renewable energy sources. For making a decision regarding development of a SHP project, its financial viability is to be assessed along with technical feasibility. In the present study, a methodology has been developed for cost assessment of high head run of river (ROR) SHP projects in order to determine their techno-economic viability before undergoing detailed investigation. This will enable the planners and developers to carry out detailed investigation and implementation of only those projects which are financially feasible. In the analysis of cost, capacity and head have been considered cost-influencing parameters. The correlations for cost based on different types of head race conduit, penstock materials, types of turbine and types of generator for various layouts have been developed. The project cost determined from the developed correlations is validated by the cost data collected from recently developed projects. It has been found that the developed correlation for cost may be used for reasonable cost estimation of hydropower projects for planning of such projects. The cost analysis of different layouts based on generation cost and installation cost were evaluated. Particle Swarm Optimization (PSO) based technique has been employed to work out the optimal layout. Based on cost analysis, nomogram has been developed for selection of optimum layout for high head RoR SHP projects.

Keywords Small hydropower (SHP) · Run of river (RoR) · Cost correlations · PSO · Nomogram

S. Mishra
School of Electrical and Electronics Engineering, Lovely Professional University, Phagwara, Punjab, India
e-mail: rite2sm@gmail.com

S.K. Singal (✉) · D.K. Khatod
Alternate Hydro Energy Centre, Indian Institute of Technology Roorkee, Roorkee, Uttarakhand, India
e-mail: sunilksingal@gmail.com

D.K. Khatod
e-mail: khatoddheeraj@gmail.com

© Springer Nature Singapore Pte Ltd. 2018
M. Majumder (ed.), *Application of Geographical Information Systems and Soft Computation Techniques in Water and Water Based Renewable Energy Problems*, Water Resources Development and Management, https://doi.org/10.1007/978-981-10-6205-6_7

7.1 Introduction

Small hydropower (SHP) is a renewable source of energy and available Worldwide. An estimated potential of about 19,749 MW from 6474 project sites exist in India [1, 2]. Installed capacities from few kW to 25 MW are covered under "Small Hydro". Based on operating head SHP sites are further divided into three categories having following ranges [3].

Low head: 2–30 m head.
Medium head: 30–100 m head.
High head: More than 100 m head.

For development of a project, cost estimate is very important in addition to technical feasibility. Many researchers have developed some correlations for cost of such projects in the past. Ogayar and Vidal [4] developed the correlation for determining the cost of electro-mechanical equipment of small hydropower plant considering power and head as basic parameters. Based on different types of turbines i.e. Semi Kaplan, Kaplan, Pelton and Francis, they developed different cost correlations. These equations were applicable up to 2 MW installed capacity. The errors were found to be within 20%. For refurbishment of SHP plant, Ogayar et al. [5] developed simple cost correlations of different elements based on the economic optimization.

In small hydropower development reduction in cost were discussed by Minott and Delisser [6] and Mishra et al. [7, 8]. They discussed that in the developing countries the capital costs could be reduced by the use of various materials such as; PVC, polyethylene, asbestos,, reinforced polyester, wood and fiber glass cement, etc. as penstock material. By the use of electronic sensors, cost reduction in speed control of turbine can be achieved rather than going for conventional method [6]. By the use of reverse pump prime movers cost reduction in case of turbines can be achieved rather than going for separate turbine and pumps [8].

Gordon [9] developed a simple methodology for estimating preliminary cost of hydropower projects. This methodology was based on the analysis of cost data of existing projects. The correlations for the cost of hydropower projects were developed with respect to head and capacity. The accuracy of such cost estimates was found to be 50%. These correlations are mostly applicable to large hydropower schemes having medium and high heads. Singal and Saini [10] developed the cost correlations of low head run of river small hydropower projects considering capacity and head as the cost-influencing parameters. The maximum deviation of ±11% was found between the actual costs and the cost predicted based on correlations.

Thus, earlier studies for the development of correlations for the estimation of cost of hydropower projects were based on cost data collected for existing projects or cost analysis for the projects under low head range. Under the present study, an attempt has been made to analyze the cost of various components of high head run of river small hydropower projects based on the actual quantities of various items and the prevailing prices of these items. Head and capacity have been identified as the cost sensitive parameters and the correlations are developed for cost estimation as function of these parameters. This study has been carried out for North and North east regions of India especially for the Himalayan ranges as most of the high head SHP schemes fall in this region.

7.2 Methodology

Civil works and electro-mechanical equipment constitute the major portion of cost of SHP scheme. The parameters on which size of civil works and electro-mechanical equipment depend are the installed capacity, runner diameter of turbine, discharge and head.

An extensive exercise has been carried out to generate the cost data. The methodology adopted for computing cost for each case is given in the following steps.

7.2.1 Generation of Cost Data for Civil Work Components

The following steps are involved in generation of cost data for civil works:

1. Various combinations of head in range of 100 to 1000 m and plant capacity in the range of 2000 to 25,000 kW were considered [10].
2. Based on the head and capacity discharge was computed.
3. Sizing of each component of civil works was carried out based on discharge and drawings of each component were prepared by hydraulic design.
4. From the drawings, quantity (volume/weight) of each item for various civil works was calculated.
5. Considering the prevailing market prices for each item, the costs of various civil works were computed.
6. The above exercise was repeated for other combinations of head and capacity considered under the present study.

7.2.2 Development of Correlations

The cost correlations obtained in the present article are in Indian rupees (INR).[1] The equation for cost per kW is shown in Eq. (7.1).

$$C = f(a, b) \qquad (7.1)$$

---

[1] 1 USD = 55 INR.
1 EUR = 71 INR.

where a, b are the cost influencing parameters. A correlation has been statistically developed for cost per kW (C) by regression analysis of the analyzed data. The methodology developed by the authors (Mishra et al. [11, 12]) has been used for the development of correlations.

7.3 Analysis for Project Cost

The cost of components for various layouts of high head RoR SHP schemes having different capacity and head have been determined based on actual quantities of different items at prevailing rates. The SHP schemes considered in the study have two units in the head range 100–1000 m and unit size 1000–12,500 kW [10]. The sizes of civil works components are determined based on discharge. Thus discharge (in m³/s) has been calculated for each layout and sizing of civil work has been done by using the methodology given in Sect. 7.2.1. For each layout, runner diameter (d) of turbine in millimeter is calculated by using the following equations (Eqs. (7.2)–(7.5)) [13] as the size of turbine is represented by its runner diameter. Based on runner diameter of the turbine, the layout of powerhouse is worked out.

$$d = \frac{84.6 \; \theta_3 (H)^{1/2}}{N} \tag{7.2}$$

$$\begin{aligned} \theta_3 &= 0.0211(N_s)^{2/3} & \textit{(for reaction turbine)} \\ \theta_3 \tfrac{d_j}{d} &= 0.0019 N_s & \textit{(for impulse turbine)} \end{aligned} \tag{7.3}$$

where, d_j is the diameter of the jet.

$$\theta_3 = 0.0342(N_S) \tag{7.4}$$

considering, normal diameter ratio as 0.0555.

$$N_S = \frac{N\sqrt{P}}{H^{5/4}} \tag{7.5}$$

where, N_s is the specific speed of turbine; N is the rotational speed of turbine in revolution per minute; H is the rated net head in m; P is the rated output power in kW at full gate opening; θ_3 is the velocity ratio at discharge diameter of runner.

7.3.1 Civil Works

The cost of various components of civil works has been determined using the methodology explained under Sect. 7.2.1 for different cases considered.

As an example, a project having a plant capacity of 5000 kW at 100 m head has been considered, sizing was carried out and the drawings were prepared. From the drawings, quantities of various items were calculated as given in Table 7.1. The important items contributing cost are earth work in excavation, concreting, reinforced steel and structural steel.

7.3.1.1 Analysis for Development of Cost Correlation

For different combinations of head and capacity, the quantities of different items have been computed by following the same methodology as discussed in Sect. 7.2 and shown in Table 7.1 for one case. Based on these data, correlation has been developed by first-order regression technique as discussed in Sect. 7.2.2. The developed correlation for the estimation of quantities of various civil works components are shown in Table 7.2.

7.3.1.2 Cost Analysis

For computation of cost of each component of civil works, the developed correlation for quantities of different items and prevailing item rates for year 2012 as shown in Table 7.3 have been used. The main items in civil works are earth work in excavation, concreting, reinforced steel and structural steel. There are some minor miscellaneous items such as doors, windows, plastering, water supply works, sanitary works, drainage, fencing and paintings etc. The cost of these miscellaneous items is about 6% of the cost of main items in each component [10]. Accordingly cost of miscellaneous items have also been computed and total cost of civil works components such as diversion weir and intake, intake channel, desilting tank, head race channel, forebay and spillway, penstock, surge tank, power house building and tail race channel has also been computed by using the Eq. (7.6).

$$C_x = 1.06 \times ((E/W \times cost\,of\,earth\,work\,in\,excavation) + (Conc. \times cost\,of\,Concerting)$$
$$+ (RS \times cost\,of\,Reinforced\,steel) + (SS \times cost\,of\,Structural\,Steel/Material))$$

$$(7.6)$$

Finally, cost per kilowatt of civil works (C_c) has been worked out using Eq. (7.7).

$$C_C = C_W + C_{IC} + C_{DT} + C_{HRC} + C_{FS} + C_P + C_{PH} + C_{TRC} \qquad (7.7)$$

7.3.2 Electro-Mechanical Equipment

Cost values of electro-mechanical equipment considering different turbines were collected from various manufacturers for a range of capacity, head, and runner

Table 7.1 Drawings and quantities of components for a project having 5000 kW capacity at 100 m head

S. No.	Component	Drawing	Quantity of various items
1	Diversion weir and intake		Earth work in excavation = 5032.43 m³ Concreting = 3680.03 m³ Reinforced steel = 285.66 MT Structural steel = 32.22 MT
2	Intake channel		Earth work in excavation = 5996.522 m³ Concreting = 1829.015 m³ Reinforced steel = 0.603 MT

(continued)

Table 7.1 (continued)

S. No.	Component	Drawing	Quantity of various items
3	Desilting tank [11]		Earth work in excavation = 6949 m^3 Concreting = 1652 m^3 Reinforced steel = 119.05 MT Structural steel = 9.27 MT

(continued)

Table 7.1 (continued)

S. No.	Component	Drawing	Quantity of various items
4	Head race channel		Earth work in excavation = 4241.19 m^3 Concreting = 1362.56 m^3 Reinforced steel = 0.07 MT
5	Forebay and spillway		Earth work in excavation = 3090.21 m^3 Concreting = 1675.02 m^3 Reinforced steel = 129.21 MT Structural steel = 23.63 MT

(continued)

Table 7.1 (continued)

S. No.	Component	Drawing	Quantity of various items
6	Penstock		Earth work in excavation = 1236.46 m^3 Concreting = 927.42 m^3 Reinforced steel = 21.6 MT Structural steel = 164.68 MT
8	Power house building		Earth work in excavation = 7457.23 m^3 Concreting = 1325.04 m^3 Reinforced steel = 165.91 MT Structural steel = 21.6 MT

(continued)

Table 7.1 (continued)

S. No.	Component	Drawing	Quantity of various items
9	Tail race channel		Earth work in excavation = 417.73 m^3 Concreting = 24.01 m^3 Reinforced steel = 0.014 MT

Table 7.2 Correlations for civil works for RoR SHP scheme

S. No.	Civil works components		Items			
			Earth work in excavation (m^3), E/W	Concreting (m^3), Conc.	Reinforcement steel (MT), RS	Structural steel/material (MT), SS
1	Diversion weir		$47.00\,P^{1.10}$ $H^{-0.99}$	$38.55\,P^{1.17}$ $H^{-1.16}$	$2.59\,P^{1.18}$ $H^{-1.15}$	$1.51\,P^{0.71}$ $H^{-0.67}$
2	Intake channel (per meter)		$2.99\,P^{0.85}$ $H^{-0.91}$	$0.81\,P^{0.88}$ $H^{-0.94}$	$0.03\,P^{0.82}$ $H^{-0.87}$	–
3	Desilting tank		$1770.70\,P^{0.83}$ $H^{-1.02}$	836.96 $P^{0.79}\,H^{-1.01}$	$9.81\,P^{0.76}$ $H^{-0.92}$	$2.96\,P^{0.83}$ $H^{-1.01}$
4	Head race (per meter)	Channel	$0.27\,P^{0.84}$ $H^{-0.94}$	$0.09\,P^{0.84}$ $H^{-0.95}$	$0.02\,P^{0.22}$ $H^{-0.25}$	–
		PVC pipe	$2.0\,P^{0.84}$ $H^{-0.95}$	$0.13\,P^{0.87}$ $H^{-0.98}$	$0.01\,P^{0.82}$ $H^{-0.92}$	$0.01\,P^{1.5}$ $H^{-0.93}$
		MS pipe				$0.02\,P^{1.10}$ $H^{-1.24}$
		GRP pipe				$0.01\,P^{1.5}$ $H^{-0.5}$
		Tunnel	$0.08\,P^{0.98}$ $H^{-0.95}$	$0.83\,P^{0.71}$ $H^{-0.69}$	–	$0.02\,P^{0.96}$ $H^{-0.93}$
5	Surge tank		$0.04\,P^{0.98}$ $H^{0.19}$	$0.04\,P^{0.98}$ $H^{0.20}$	$0.03\,P^{0.98}\,H^{0.19}$	–
6	Forebay and spillway		$1339.43\,P^{0.50}$ $H^{-0.69}$	$70.31\,P^{0.73}$ $H^{-0.63}$	$5.22\,P^{0.72}$ $H^{-0.61}$	$5.62\,P^{0.58}$ $H^{-0.70}$
7	Penstock (per meter)	PVC pipe	$0.42\,P^{0.83}$ $H^{-0.98}$	$0.31\,P^{0.84}$ $H^{-0.98}$	$0.03\,P^{0.83}$ $H^{-0.97}$	$0.04\,P^{1.5}$ $H^{-0.81}$
		GRP pipe				$0.01\,P^{1.5}$ $H^{-0.85}$
		HDPE pipe				$0.08\,P^{1.5}$ $H^{-0.80}$
		Steel pipe				$0.05\,P^{0.83}$ $H^{-0.95}$
8	Tail race channel (per meter)		$0.91\,P^{0.83}$ $H^{-0.90}$	$1.82\,P^{0.87}$ $H^{-0.91}$	$0.01\,P^{0.80}$ $H^{-0.79}$	–
9	Power house building	Pelton/turgo impulse	$16.09\,P^{2.46}$ $H^{-1.75}$	0.00052 $P^{2.54}\,H^{-0.42}$	$0.00022\,P^{4.02}$ $H^{-2.83}$	$0.09\,P^{4.55}$ $H^{-3.50}$
		Francis	$0.08\,P^{2.33}$ $H^{-1.33}$	$0.03\,P^{2.29}$ $H^{-1.32}$	$0.05\,P^{2.36}$ $H^{-1.34}$	$0.02\,P^{2.19}$ $H^{-1.26}$

diameter considered under the present study. Same methodology as mentioned in Sect. 7.2 has been employed for the development of cost correlation for electro-mechanical equipment considering different turbines. Therefore, the cost correlation developed for the different types of electromechanical equipment, C_y is shown in Eq. (7.8).

Table 7.3 Price as per schedule of rates prevailing for the year 2012

S. No.	Items	Price (INR)
1	Earthwork in excavation with all leads and lifts (a) In ordinary soil (b) In soft rock, where blasting is not required (c) In hard rock including blasting	$265/m^3$ $330/m^3$ $550/m^3$
2	M20 grade concrete work in plain cement concrete as well as in reinforced cement concrete including shuttering, mixing, placing in position, compacting, and curing	$3640/m^3$
3	Reinforcement steel bars of *iron* 500 grade including cutting, bending, binding, and placing in position	55,000/MT
4	Structural steel including fabrication, transportation to site, and erection	75,000/MT

Table 7.4 Coefficients in cost correlation for electro-mechanical equipment with different types of turbines [12]

S. No.	Type of equipment		Coefficients in cost correlation		
			a_1	x_1	x_2
1	Turbine with governing system (TG)	Pelton	117,313	−0.03	−0.39
		Turgo impulse	145,121	−0.12	−0.24
		Francis	125,354	−0.01	−0.38
2	Generator with excitation system or capacitor bank (GE)	Induction	130,262	−0.19	−0.22
		Synchronous	143,660	−0.18	−0.21
3	Auxiliaries		21,846	−0.19	−0.22
4	Transformer		221	0.11	0.01
5	Switchyard		1.82	0.17	0.93

Cost correlation for different electromechanical equipment, $C_y = a_1 P^{x_1} H^{x_2}$ (7.8)

The value of the constants a_1, x_1 and x_2 shown in Eq. (7.8) for cost of electro-mechanical equipment as determined for different type of turbine are given in Table 7.4. The cost per kW of electromechanical equipment is given by Eq. (7.9).

$$C_{e\&m} = C_{TG} + C_{GE} + C_{Aux} + C_{T/F} + C_{SY} \tag{7.9}$$

7.3.3 Total Cost

The total project cost includes the cost of civil works, the cost of electro-mechanical equipment and the cost of various items and other indirect costs. Miscellaneous and indirect cost includes the costs of designs, indirect costs, tools and plants,

Fig. 7.1 Cost per kW of
RoR SHP scheme

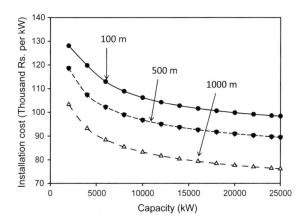

communication costs, preliminary charge of preparing the report, survey and
investigation, environmental impact assessment and cost of land. The cost of these
miscellaneous items comes out to 13% the sum of the cost of civil works and
electro-mechanical equipment. The cost of the operation and maintenance is considered to be about 1.5% of the total project cost [10, 14]. The total cost per kW
(C) of the project as shown in the Eq. (7.10) is the sum of the cost per kW of civil
works, as shown in the Eq. (7.7) and the cost per kW of electro-mechanical
equipment as shown in the Eq. (7.9) and 13% of the sum of the cost of civil works
and electro-mechanical equipment. Figure 7.1 shows the trend of the installation
cost at different heads.

$$C = 1.13(C_C + C_{e\&m}) \tag{7.10}$$

7.4 Cost Optimization

A financial analysis has been carried out for cost optimization based on minimum
generation cost (G).

7.4.1 Objective Function

The objective function of the financial analysis of SHP projects is to minimize the
generation cost (G) of the SHP project. It can be expressed mathematically as
shown in Eq. (7.11). The problem constraints in Eqs. (7.18–7.26) are the optimized
parameter bounds. Therefore, the design problem can be formulated as the following optimization problem:

$$\text{Minimize}, G = C_{\exp}/E \tag{7.11}$$

where, $C_{\exp}$ is the annual expenditure and is given by:

$$C_{\exp} = (l \times C_f \times \frac{o\&m}{100}) + (l \times C_f \times \frac{dep}{100}) + (l \times C_f \times \frac{IR}{100}) \tag{7.12}$$

The value of *o&m*, *dep*, *IR* and *l* taken for calculation has been given in Table 7.5.

$$E = 8760 \times P \times (1 - \frac{H_{loss}}{H_{gross}}) \times (\frac{\eta_t}{\eta_f}) \times plf \tag{7.13}$$

$$C_f = C + \sum_{i=0}^{n_i-1} [x_i \times IR \times l \times C \times (yr - (i \times \frac{yr}{n_i})) \tag{7.14}$$

$$C = (C_C + C_{e\&m}) + 0.13(C_C + C_{e\&m}) \tag{7.15}$$

$$C_C = C_W + C_{IC} + C_{DT} + \sum_{i=1}^{N_h} X_i^h C_{HR}(i) + \sum_{i=1}^{N_p} X_i^p C_P(i) + C_{PH} + C_{TRC} \tag{7.16}$$

$$C_{e\&m} = C_{oe\&m} + \sum_{i=1}^{N_t} X_i^t C_t(i) + \sum_{i=1}^{N_g} C_g(i) X_i^g \tag{7.17}$$

With $C_{oe\&m} = C_{Aux} + C_{T/F} + C_{SY}$

Based on technical feasibility, different alternatives considered for optimization are given in Table 7.5. The technically feasible types of turbines under the head

Table 7.5 Alternatives considered for optimization

S. No.	Components	Alternatives
1	Head race conduit	Channel
		Glass reinforced plastic (GRP) pipe
		Mild steel (MS) pipe
		Tunnel
2	Penstock	Poly vinyl chloride (PVC) pipe
		Glass reinforced plastic (GRP) pipe
		High density poly ethylene (HDPE) pipe
		Steel pipe
3	Turbine	Francis
		Turgo impulse
		Pelton
4	Generator	Induction
		Synchronous

range considered are Francis, Turgo Impulse and Pelton. Whereas, penstock material considered are steel, poly vinyl chloride (PVC), high density poly ethylene (HDPE) and glass reinforced plastic (GRP).

Where, X_i^h represent the options considered for headrace conduits i.e. channel, pipe material (such as PVC and mild steel) and tunnel (where, $i = 1$ for head race channel, 2 for head race PVC pipe, 3 for head race GRP pipe, 4 for head race MS pipe, 5 for head race tunnel). X_i^p represents the option of material considered for penstock (such as PVC, GRP, HDPE and steel) (where, $i = 1$ for PVC pipe, 2 for HDPE pipe, 3 for GRP pipe, 4 for steel pipe). X_i^t represents the different types of turbines (where, $i = 1$ for Francis turbine, 2 for Pelton turbine, 3 for turgo impulse) and X_i^g represent the different types of generator applicable for high head SHP projects (where, $i = 1$ for induction generator with capacitor bank, 2 for synchronous generator with excitation system). N_h, N_p, N_t and N_g are the number of available options for headrace conduit, penstock material, turbine type and generator type respectively. $C_{HR}(i)$, $C_P(i)$, $C_t(i)$ and $C_g(i)$ are the cost associated with the ith option of the headrace conduit, penstock material, turbine type and generator type respectively (Table 7.6).

$$\text{Subject to,} \sum_{i=1}^{N_h} X_i^h = 1 \tag{7.18}$$

$$\sum_{i=1}^{N_p} X_i^p = 1 \tag{7.19}$$

$$\sum_{i=1}^{N_t} X_i^t = 1 \tag{7.20}$$

$$\sum_{i=1}^{N_g} X_i^g = 1 \tag{7.21}$$

$$X_1^g = \begin{cases} 1 & if \quad P \le 10MW \\ 0 & otherwise \end{cases} \tag{7.22}$$

Table 7.6 Values of parameters considered for financial analysis

S. No.	Parameter	Value
1	Annual interest rate (IR)	11%
2	Annual depreciation (dep.)	3.4%
3	Annual operation and maintenance cost (o&m)	1.5%
4	Life of plant considered for analysis	35 years
5	Construction period	3 years
6	Debt equity ratio	70:30

$$X_1^P = \begin{matrix} 1 \ if \ P \leq 2.5 \ MW \& H \leq 150m \\ 0 \quad otherwise \end{matrix} \tag{7.23}$$

$$X_2^P = \begin{cases} 1 & if \ 2.5MW < P \leq 5MW \& 100 \leq H \leq 650m \\ 0 & otherwise \end{cases} \tag{7.24}$$

$$X_3^P = \begin{cases} 1 & if \ 2MW < P \leq 25MW \& 100 \leq H \leq 750m \\ 0 & otherwise \end{cases} \tag{7.25}$$

$$\begin{aligned} N^h &= 5 \\ N^p &= 4 \\ N^t &= 3 \\ N^g &= 2 \end{aligned} \tag{7.26}$$

7.4.2 Particle Swarm Optimization Technique

One of the evolutionary computation technique whose search method is based on natural system is Particle Swarm Optimization (PSO), which was developed by Kennedy and Eberhart. Like genetic algorithm (GA), PSO is also a population (swarm) based optimization tool. In case of PSO crossover and mutation are carried out simultaneously whereas, in GA these process are carried out consecutively. One major difference between particle swarm and traditional evolutionary computation methods is that particles' velocities are adjusted, while evolutionary individuals' positions are acted upon; it is as if the "fate" is altered rather than the "state" of the particle swarm individuals [15].

The PSO technique can generate high quality solutions in reducing the computation time and stable convergence characteristic than other stochastic methods [16]. Although the PSO seems to be sensitive to the adjustment of certain weight or parameters, many studies are still underway to prove its potential to solve the problems of complex power system [16–18]. It was found that the PSO method quickly finds the high quality solution for many optimization problems of power system. However, PSO has a capacity of more global search at the beginning and a local search at the end. Therefore, while solving problems with more local optima, there are more possibilities for the PSO to explore local optima at the end of the run.

7.4.3   Optimization Methodologies for Generation Cost Problems

Algorithms based on PSO search in parallel using a team of individuals similar to other optimization techniques based on evolutionary algorithms. Each individual is a candidate solution to the problem. Individuals in a swarm approach reaches to its optimum value through its current speed, experience, and the experience of its neighbors. In a search space of n physical dimensions, the position and the speed of the individual i at kth iteration are represented as the vectors $X_i^k = \left(x_{i1}^k, x_{i2}^k, \ldots, x_{in}^k\right)$ and $V_i^k = \left(v_{i1}^k, v_{i2}^k, \ldots, v_{in}^k\right)$ respectively in the PSO algorithm. Let $Pbest_i^k = \left(x_{i1}^{Pbest}, x_{i2}^{Pbest}, \ldots, x_{in}^{Pbest}\right)$ and $Gbest^k = \left(x_1^{Pbest}, x_2^{Pbest}, \ldots, x_n^{Pbest}\right)$ be the best position of individual i and its neighbors' best position till kth iteration respectively. Using this information, the updated velocity of individual i is modified during $k + 1$th iteration as given in Eq. (7.27) [19]:

$$V_i^{k+1} = wV_i^k + c_1r_1 \times \left(Pbest_i^k - X_i^k\right) + c_2r_2 \times \left(Gbest^k - X_i^k\right) \quad (7.27)$$

where,

V_i^k	Velocity of individual i at iteration k.
W	Inertia weight parameter.
c_1, c_2	Acceleration coefficients i.e. $c_1 = 2$ and $c_2 = 4$.
r_1, r_2	Random number between 0 and 1.
X_i^k	Position of individual i at iteration k.
$Pbest_i^k$	Best position of individual i till iteration k.
$Gbest^k$	Best position of the group till iteration k.

In this velocity updating process, the values of parameters such as w, c_1 and c_2 are determined in advance. In general, the weight w is set according to Eq. (7.28).

$$w = w_{max} - \frac{w_{max} - w_{min}}{iter_{max}} \cdot iter \quad (7.28)$$

where,

w_{min}, w_{max}	Minimum and maximum weights i.e. $w_{max} = 0.9$ and $w_{min} = 0.4$
$iter_{max}$	Maximum iteration number
$Iter$	Current iteration number.

Each individual moves from the current position to the next one by the modified velocity in Eq. (7.27) using (7.28).

$$X_i^{k+1} = X_i^k + V_i^{k+1} \quad (7.29)$$

Figure 7.2 shows the process of finding the optimal solution by PSO technique.

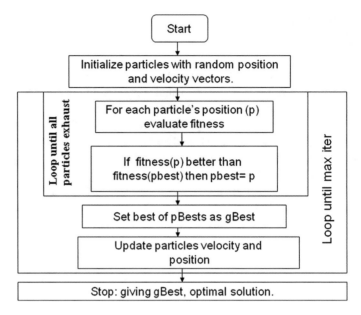

Fig. 7.2 Flowchart of PSO based technique

7.5 Results and Discussions

An analysis for the development of correlations for the cost of various components of high head run of river SHP scheme having two units of generating machine in the head range 100–1000 m and unit size 1000–12,500 kW has been carried out. By regression analysis, correlations for quantities of various items required for civil works are developed. Similarly, by regression analysis, correlation for cost per kW of electro-mechanical equipment is also developed. Penstock, power channel and head race channel are more site specific because their costs depend on their lengths and topography. For similar head and capacity, the length of channel and penstock may vary in case country slope of sites are different. Considering this fact, quantities required for construction of the channel and the penstocks have been estimated as per unit length based on capacity and head. It is seen that the cost of channel and penstock increases with head. This is in order on the line that, for an increasing head, the length will be more, hence the cost. The cost of SHP plants as determined from developed correlations based on actual quantities of various items and prevailing rates has been compared with the cost data of existing/planned SHP plants. For this, the actual cost data from 32 small hydropower plants installed/planned during last 10 years were collected as given in Appendix 1. The cost data of these plants pertains to different periods. Therefore, the collected cost data have been escalated based on consumer price index to bring all the costs at the level of base year 2012. Figure 7.3 shows the comparison of cost per kW of civil

Fig. 7.3 Comparison of cost per kW of civil works as analyzed with collected project data

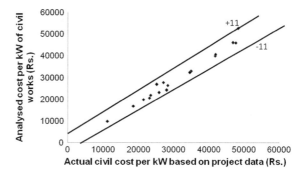

Fig. 7.4 Comparison of cost per kW of E&M equipment as analyzed with collected project data

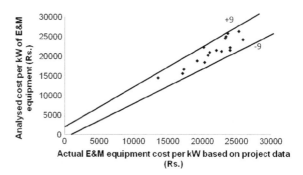

Fig. 7.5 Comparison of total cost per kW as analyzed with collected project data

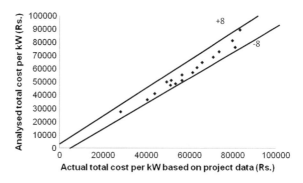

works, where maximum deviation of ±11% has been found, however 94% of the cost data are within ±6%. Comparison of costs for electro-mechanical equipment is shown in Fig. 7.4. It is seen from the Fig. 7.4 that a maximum deviation for these costs is ±9%. Similarly, a comparison of total cost shown in Fig. 7.5 depicts that 93% of the data are within ±4%, which shows that the accuracy of correlation is even better than the maximum deviation of ±8%.

PSO based technique has been employed for the optimization of generation cost of various layouts of RoR based SHP scheme. To determine the optimum layout, RoR SHP schemes having different types of turbine, types of generator, penstock material and head race conduit are considered taking turbine speed as 1000 rpm and plant load factor (*plf*) as 0.60. The values considered for the constants used in PSO based optimization technique are: population size = 10, maximum iteration = 100, $c_1 = 2$, $c_2 = 1.5$, $w_{min} = 0.4$, $w_{max} = 0.9$. Three different projects having plant capacity of 2 MW, 5 MW and 15 MW with head ranging from 100 to 1000 m has been considered to obtain the optimal layout of the RoR SHP scheme. The optimal selection of components in layout of the projects is given in Tables 7.7, 7.8 and 7.9. The convergence of the PSO technique for optimization to reach the optimal generation cost for the above mentioned projects is shown in Fig. 7.6. Similar convergence of the PSO technique for optimization for different combinations were also obtained. The graphical representation of the optimal layout with generation cost is shown in Figs. 7.7, 7.8 and 7.9.

Table 7.7 Optimal layout of 2000 kW RoR SHP scheme

S. No.	Head (m)	Generation cost (Rs./kWh)	Head race	Penstock	Turbine	Generator
1	100	3.88	Channel	PVC	Francis	Induction
2	200	4.61	Channel	HDPE	Turgo impulse	Induction
3	300	3.69	Channel	HDPE	Pelton	Induction
4	400	3.31	Channel	HDPE	Pelton	Induction
5	500	3.07	Channel	HDPE	Pelton	Induction
6	600	2.90	Channel	HDPE	Pelton	Induction
7	700	2.96	Channel	GRP	Pelton (2 nozzle)	Induction
8	800	2.83	Channel	GRP	Pelton (2 nozzle)	Induction
9	900	3.31	Channel	Steel	Pelton (3 nozzle)	Induction
10	1000	3.41	Channel	Steel	Pelton (4 nozzle)	Induction

Table 7.8 Optimal layout of 5000 kW RoR SHP scheme

S. No.	Head (m)	Generation cost (Rs./kWh)	Head Race	Penstock	Turbine	Generator
1	100	2.94	Channel	HDPE	Francis	Induction
2	200	3.71	Channel	HDPE	Turgo impulse	Induction
3	300	3.13	Channel	HDPE	Turgo impulse	Induction
4	400	2.74	Channel	HDPE	Pelton	Induction
5	500	2.55	Channel	HDPE	Pelton	Induction
6	600	2.42	Channel	HDPE	Pelton	Induction
7	700	2.22	Channel	GRP	Pelton	Induction
8	800	2.14	Channel	GRP	Pelton	Induction
9	900	2.31	Channel	Steel	Pelton	Induction
10	1000	2.46	Channel	Steel	Pelton (2 nozzle)	Induction

Table 7.9 Optimal layout of 15,000 kW RoR SHP scheme

S. No.	Head (m)	Generation cost (Rs./kWh)	Head race	Penstock	Turbine	Generator
1	100	2.38	Channel	GRP	Francis	Synchronous
2	200	2.08	Channel	GRP	Francis	Synchronous
3	300	2.64	Channel	GRP	Turgo impulse	Synchronous
4	400	2.40	Channel	GRP	Turgo impulse	Synchronous
5	500	2.25	Channel	GRP	Turgo impulse	Synchronous
6	600	2.19	Channel	GRP	Pelton	Synchronous
7	700	2.12	Channel	GRP	Pelton	Synchronous
8	800	2.04	Channel	GRP	Pelton	Synchronous
9	900	2.15	Channel	Steel	Pelton	Synchronous
10	1000	2.10	Channel	Steel	Pelton	Synchronous

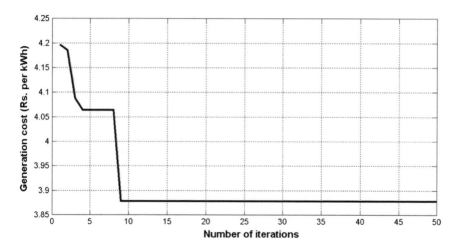

Fig. 7.6 Convergence of optimization problem using PSO based technique for 2 MW capacity at 100 m head

Fig. 7.7 Optimal layout of 2 MW capacity at different heads

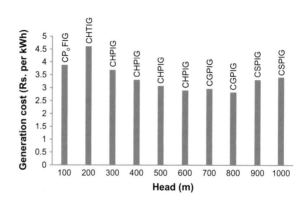

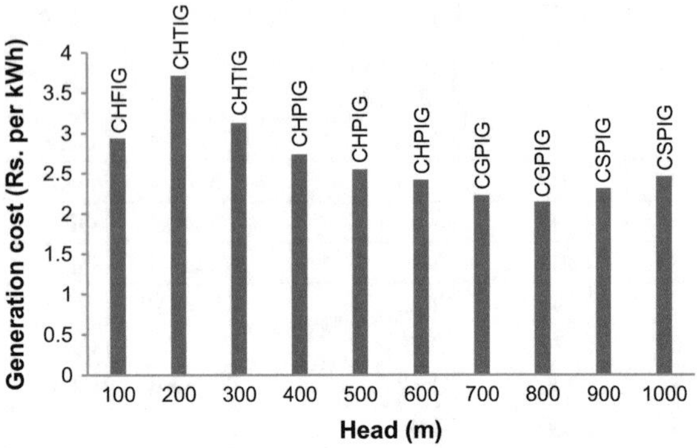

Fig. 7.8 Optimal layout of 5 MW capacity at different heads. *C* Channel, *P_o* PVC penstock, *H* HDPE penstock, *G* GRP penstock, *S* Steel penstock, *IG* Induction generator with capacitor bank, *SG* Synchronous generator with excitation system, *F* Francis turbine, *T* Turgo impulse, *P* Pelton turbine

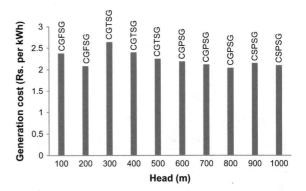

Fig. 7.9 Optimal layout of 15 MW capacity at different heads

From Fig. 7.6, it is clear that the optimization problem converges well and is similar in all the other cases. Also, in most of the cases, it reaches to the optimal solution within 10 iterations. This shows the appropriateness of developed formulation and PSO based technique. Figures 7.7, 7.8 and 7.9 shows the optimum selection of components at different heads and capacities. Figure 7.7 shows that the optimum layout for 2 MW capacity SHP project at 100 m head have head race channel for head race conduit, PVC pipe for penstock, Francis turbine and induction generator. Figure 7.8 shows the optimum layout for 5 MW capacity SHP project at 500 m head with head race channel for head race conduit, HDPE pipe for penstock, Pelton turbine and induction generator. As seen from Fig. 7.9, 15 MW capacity project at 1000 m head have optimum layout with head race channel, steel pipe for penstock, Pelton turbine and synchronous generator.

Further an attempt has been made to develop the nomograms as shown in Fig. 7.10. Figure 7.10 shows the nomogram for optimum selection of different components considered for optimum layout in high head SHP scheme. It is seen that PVC pipe for penstock, channel for head race conduit, Francis turbine and induction generator is the optimum selection for head range of 100–150 m and capacity up to 5 MW. Whereas, HDPE pipe for penstock, channel for head race conduit, Turgo Impulse turbine and induction generator is the optimum selection

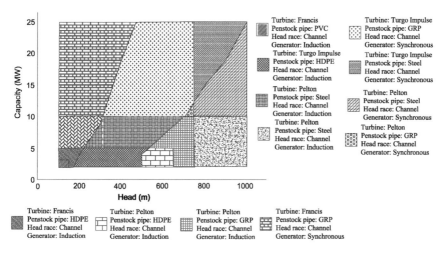

Fig. 7.10 Nomogram for selection of component. *C* Channel, *P_o* PVC penstock, *H* HDPE penstock, *G* GRP penstock, *S* Steel penstock, *IG* Induction generator with capacitor bank, *SG* Synchronous generator with excitation system, *F* Francis turbine, *T* Turgo impulse, *P* Pelton turbine

for head range of 200–550 m and capacity up to 5 MW. While steel pipe, head race channel, Pelton turbine and Induction generator is optimum selection for a range of 200–650 m and capacity range of 5–10 MW. Pelton turbine, steel penstock, head race channel and synchronous generator are the optimum layout for the head range of 750–1000 m and capacity range of 10–25 MW.

7.6 Conclusions

An attempt has been made to analyze the cost of the high head RoR SHP schemes and the correlations for estimating the quantities of items of various civil works are developed. The correlations are developed by regression analysis based on actual quantities of various items and costs are determined by multiplying the quantities of items with the prevailing rates. It is found that civil works constitute the major portion of the overall cost for a high-head scheme which increases with increase in head. The accuracy of the correlations has been validated by comparing with cost data collected for existing power stations commissioned recently given in Annexure I. A maximum deviation of ±8% has been found, however 94% cost data are within ±4%. Therefore these co-relations can be taken as basis for determining the cost of the high head RoR SHP schemes. A PSO based technique has been used to optimize the generation cost of a project in order to determine the optimal layout. This exercise has been repeated for different combinations of head and installed capacity. Based on the optimal layout obtained, the nomogram has been developed. The methodology employed for determination of the optimum installation and the nomograms developed for selection of optimum layouts of high head SHP schemes can be used by developers, policy makers and decision takers to plan their investments in development of small hydropower projects in North and North east regions of India especially for the Himalayan ranges. The financers may also use these correlations for appraisal of such schemes for financing. In this study, two generating units were considered. For future research, this work can be extended by considering more than two units. An assumption has also been considered that the penstock of the small hydropower project is straight and each unit has a single penstock coming from the forebay tank. Future researchers may consider a single penstock coming from forebay and getting branched for each machine.

Acknowledgements The authors acknowledge the use of project data as available in AHEC, IIT Roorkee.

Appendix-1

Cost Data Collected from Existing SHP Plants (Base Year 2012)

S. No.	Name of the project	Head (m)	Capacity (kW)	Actual project cost (Rs./kW)		
				Civil works	Electro-mechanical equipment	Total installation cost
1	Manglay	118	2000	46,846	23,721	79,741
2	Hanumanganga	150	3000	48,294	25,309	83,172
3	Sirnyuk	137	3000	41,748	23,372	73,585
4	Kakora	393	12,500	21,431	17,320	43,789
5	Vanala	135	15,000	25,821	24,047	56,351
6	Binwa Nagar	233	6000	44,638	28,969	83,176
7	Gummer	183	3000	25,048	16,541	46,996
8	Dirang	125	2000	96,486	37,408	151,300
9	Nuranang	215	6000	40,594	27,973	77,481
10	Sippi	277	4000	53,495	27,814	91,879
11	Shyam RBC	170	15,000	19,486	17,089	41,330
12	Sundarpur	130	12,000	22,530	19,651	47,666
13	Ganderbal	137	15,000	21,489	19,420	46,227
14	Mohar	120	9000	24,209	20,095	50,065
15	Branwar	220	7500	38,799	27,361	74,760
16	Athwatto	145	10,000	51,790	38,804	102,372
17	Dumkhar	157	2500	72,735	32,942	119,415
18	Kau-Tlabung	200	3000	59,192	29,508	100,232
19	Tuipanglui	243	3000	52,644	26,660	89,614

(continued)

(continued)

S. No.	Name of the project	Head (m)	Capacity (kW)	Actual project cost (Rs./kW)		
				Civil works	Electro-mechanical equipment	Total installation cost
20	Kanchauti	415	2000	43,681	19,461	71,351
21	Urgam	205	3000	58,303	29,131	98,801
22	Little Rangel	245	12,000	58,813	25,613	95,402
23	Lodhama	228	3000	54,673	27,560	92,924
24	Mungpoo Kalikhola	685	3000	32,313	16,293	54,926
25	Rimchington	293	2000	52,931	23,333	86,178
26	Dundhava	133	12,000	46,916	45,809	104,780
27	Dumartoli	167	10,000	45,312	35,030	90,788
28	Gullu—II	183	6000	35,511	30,201	74,254
29	Doe—Ama	176	13,000	39,464	32,285	81,077
30	Ataria	157	15,000	39,090	34,970	83,689
31	Sitarewa	209	10,000	30,541	21,852	59,205
32	Chittar Dam	341	2000	46,339	63,314	123,908

References

1. www.mnre.gov.in. Accessed on 15 Feb 2012.
2. www.cea.nic.in. Accessed on 15 Feb 2012.
3. European Small Hydropower Association (ESHA) (2006). Small hydropower for developing countries.
4. Ogayar, B., & Vidal, P. G. (2009). Cost determination of the electro-mechanical equipment of a small hydro-power plant. *Renewable Energy, 34,* 6–13.
5. Ogayar, B., Vidal, P. G., & Hernandez, J. C. (2009). Analysis of the cost for the refurbishment of small hydropower plants. *Renewable Energy, 34,* 2501–2509.
6. Minott, D., & Delisser, R. (1983). Cost reduction considerations in small hydropower development. In *UNIDO, Third Workshop on Small Hydropower.* Kuala Lampur, Malaysia.
7. Mishra, S., Singal, S. K., & Khatod, D. K. (2011). Approach for cost determination of electro-mechanical equipment in RoR SHP projects. *Smart Grid and Renewable Energy, 2,* 63–67.
8. Mishra, S., Singal, S. K., & Khatod, D. K. (2011). Optimal installation of small hydropower plant—A review. *Renewable and Sustainable Energy Review., 15,* 3862–3869.
9. Gordan, J. L. (1983). Hydropower cost estimates. *International Water Power & Dam Construction, 35,* 30–37.
10. Singal, S. K., & Saini, R. P. (2008). Analytical approach for development of correlations for cost of canal-based SHP schemes. *Renewable Energy, 33,* 2549–2558.
11. Mishra, S., Singal, S. K., & Khatod, D. K. (2013). Sizing and quantity estimation for desilting tank of small hydropower projects—An analytical approach. *International Journal of Green Energy, 10,* 574–586.
12. Mishra, S., Singal, S. K., & Khatod, D. K. (2013). Cost correlation for electro-mechanical equipment in small hydropower projects. *Journal of Green Energy, 10*(8), 835–847.

13. IS: 12800 (Part 3) (1991). Guidelines for the selection of hydraulic turbine, preliminary dimensioning and layout of surface hydro-electric power house (small, mini and micro hydro-electric power house). India: Bureau of Indian Standards.
14. Singal, S.K. (2008). Optimization of low head small hydropower installations. Ph.D. Thesis, Indian Institute of Technology (IIT) Roorkee.
15. Kennedy, J., & Eberhart, R. (1995). Particle swarm optimization. *IEEE International Conference Neural Networks, 4,* 1942–1948.
16. Gaing, Z. L. (2003). Particle swarm optimization to solving the economic dispatch considering the generator constraints. *IEEE Trans Power Systems, 18,* 1187–1195.
17. Yoshida, H., Kawata, K., & Fukuyama, Y. (2000). A particle swarm optimization for reactive power and voltage control considering voltage security assessment. *IEEE Trans Power Systems, 15,* 1232–1239.
18. Abido, M. A. (2002). Optimal design of power-system stabilizers using particle swarm optimization. *IEEE Transaction on Energy Conversion, 17,* 406–413.
19. Bae, P. J, Won, Jeong, Y., Houng Kim, H., & Rin Shin J. (2006). An improved particle swarm optimization for economic dispatch with value point effect. *International Journal of Innovations in Energy Systems and Power. 1,* 1–7.

Chapter 8
An Approach for Wind-Pumped Storage Plant Scheduling Under Uncertainty

J. Dhillon, A. Kumar and S.K. Singal

Abstract The combined operation of wind-pumped storage plant (PSP) system under day-ahead market has been carried out by applying stochastic technique considering uncertainty in the wind data. Such uncertainty causes risk to the scheduling of wind-PSP system. The orthogonal arrays based Taguchi method is used to manage the risk during the scheduling of wind-PSP system by maximizing the total revenue, where the numbers of experiments are performed in order to access the criticality of possible outcomes. In this study, an adjustable speed type PSP unit is considered for better reliability.

Keywords Wind system · Pumped storage plant · Optimal scheduling · Taguchi method · Uncertainty

Nomenclature

V_i	Wind speed, m/s
V_{max}, V_{min}	Maximum and minimum wind speed, m/s
S_i	Wind scenario of ith level
N_s	Total number of scenario
B	Total number of variation level across each scenario
h	Total number of testing level
V_{ci}, V_{co}	Cut-in and cut-out speed of wind turbine, m/s
V_r	Rated speed of wind turbine, m/s
P_w	Output power of wind generator, MW
P_r	Rated power of wind generator, MW
$\pi(t, s)$	Probability of sth scenarios at tth time
Rs.	Indian rupees
$R_{psp}(t, s)$	Pumped storage plant (PSP) revenue during sth scenario at tth time, Rs.

J. Dhillon (✉) · A. Kumar · S.K. Singal
Alternate Hydro Energy Centre, Indian Institute of Technology Roorkee,
Roorkee 247667, Uttarakhand, India
e-mail: javeddah@gmail.com

© Springer Nature Singapore Pte Ltd. 2018   167
M. Majumder (ed.), *Application of Geographical Information Systems and Soft Computation Techniques in Water and Water Based Renewable Energy Problems*, Water Resources Development and Management, https://doi.org/10.1007/978-981-10-6205-6_8

$R_{wind}(t,s)$	Wind farm revenue during sth scenario at tth time, Rs.
$R_{loss}(t,s)$	Revenue loss during sth scenario at tth time, Rs.
$R_{gen}(t,s)$	PSP's revenue during generating mode during sth scenario at tth time, Rs.
$R_{pump}(t,s)$	PSP's revenue during pumping mode during sth scenario at tth time, Rs.
$Pw_{pump}(t,s)$	Power supplied by wind for pumping operation during sth scenario at tth time, MW
$Pg_{pump}(t,s)$	Power supplied by grid for pumping operation during sth scenario at tth time, MW
$Pw_{mkt}(t,s)$	Power supplied by wind for market operation during sth scenario at tth time, MW
$P_{gen}(t,s)$	Power generated by PSP during sth scenario at t-th time, MW
$Pw_{unutilized}(t,s)$	Unutilized power of wind during sth scenario at tth time, MW
$Pw_{gen}(t,s)$	Power generated by wind during sth Scenario at tth time, MW
$Pp(t,s)$	Power consumed by PSP during sth scenario at tth time, MW
$E(t,s)$	Energy level of reservoir during sth scenario at tth time, MW
$Pg_{pump}^{min}, Pg_{pump}^{max}$	Minimum and maximum limit of power supplied by grid, MW
$P_{gen}^{min}, P_{gen}^{max}$	Minimum and maximum limit of power generate by PSP, MW
$P_{pump}^{min}, P_{pump}^{max}$	Minimum and maximum limit of power consumed by PSP, MW
E_{min}, E_{max}	Minimum and maximum limit of energy level in reservoir, MW
$Pd(t)$	Power demand across market, MW
$s_g(t,s), s_p(t,s)$	PSP's switching variable for generating and pumping mode
$\lambda_{mkt}(t)$	Market price, Rs./MW
$\lambda_{wind}(t)$	Wind power price, Rs./MW
ω	Penalty factor

8.1 Introduction

The integration of wind power system with the market is the biggest challenge due to its uncertain and intermittent nature, which reduces the market reliability, dispatch ability and security. Wind uncertainty or uncertainty in the wind data represents the variation in wind data in the form of wind speed. This uncertain and intermittent nature of wind resources also referred as wind uncertainty. For management of the power system, the wind uncertainty in the wind data is required to be forecasted, so that power utilities schedule their generation accordingly.

Earlier researchers analyzed number of forecasting techniques to predict the wind uncertainty, though in most of the cases, forecasting results were not accurate and brought risks in the system. To manage these risks properly, a suitable model is required. An economic dispatch problem has been evaluated by Li and Jiang for wind and thermal power systems. They studied two methods Value-At-Risk (VaR) and Integrated Risk Management (IRM) for evaluating risks in the wind energy systems [1, 2]. Abreu et al. [3] studied the combined operation of wind-hydro system in day-ahead market and used the risk constrained model to reduce the expected downside risk across the wind-hydro system. Catalao et al. [4] used two stage stochastic programming approach to maximize the profit from the wind power generation at a given level of risk, which has been evaluated using conditional value-at-risk (CVar) methodology. In the proposed study, Taguchi method is employed to calculate and manage the risks properly under uncertain condition. This experimental based design technique have the best use when there is an intermediate number of variables (3–50), having few interactions between them and only few variables contribute significantly. This method utilizes orthogonal array based structure, which is easy to implement and requires only limited number of experiments thus demands less computational time.

Due to competitive nature of the deregulated market, the uncertainty in the wind power can be reduced by integrating it with the energy storage system [5–7] such as flywheel, battery and compressed air storage etc. Energy storage technologies provide flexible, efficient, and cost effective solution to the power market to meet the challenges of integrating large-scale renewable generation into power systems. Intermittent renewable power (such as wind) can be stored in storage and released when it is needed. The benefits for such storage are lowering overall energy costs by enhancing the ability of the system to absorb generation and displace low load factor backup generation with low efficiencies, maximum use of indigenous renewable energy resources to increase the revenue across wind system, greater energy security through reduced reliance on imported fossil fuels, and a significant reduction in the carbon footprint of power systems through more sustainable energy generation and management [8–12]. For the large size wind farms, pumped storage plant (PSP) is a mature and economically viable energy storage option. Caralis et al. [13] studied the integration of wind energy with PSP. The unutilized wind energy is converted into water storage by pumping the water from lower reservoir to upper reservoir. Conversion of stored energy in upper reservoir into electrical energy is carried out effectively and efficiently. In PSP system, turbine and pump are connected separately or combined as a reversible turbine, which operates at variable speed to provide the flexibility in pumping operation. Due to variable speed operation, large quantity of wind energy is used for pumping to store the water. Due to the large uncertainty in wind energy and market price, a two stage stochastic technique has been used by Garcia-Gonzalez to make optimal decision under these uncertainties [14]. The first stage decision is the hourly bids submitted in the day-ahead market and the second stage decision is related to the operation of the pumped-storage plant for each possible realization of the random variables i.e. wind speed and market price. Angarita et al. [15] presented a stochastic approach to maximize the profit of wind-hydro system taking into account the uncertainty of wind power prediction. This mixed-integer type problem was considered to decide the bidding strategy for pool-based electricity market.

In the present study, a mixed-integer type problem has been formulated to obtain the optimal wind-PSP scheduling under uncertainty. The objective of the approach is to maximize the expected profit across wind-PSP system at the maximum level of the risk. Novelty of this work lies in the approach of scenarios selection using Taguchi method to identify the scenarios with maximum risk and optimize the system for selected set of scenarios for maximizing the overall profit. Variable operation of Pumped storage plant has also been scheduled with wind system in order to reduce the overall uncertainty in wind generation due to uncertain input.

8.2 Wind Forecast Model

To utilize the wind energy resource, spatial and temporal distributions need to be analyzed for the proper selection of wind turbine capacity. The analysis is also required to forecast and utilize the produced electricity by day-ahead market. It is important to study the characteristics of the wind frequency distribution. To fit the distribution of wind speed frequency, the Pearson model, the Rayleigh model, the Weibull model are usually applied. The Weibull distribution is often used to describe the characteristics of wind speed frequency because it is simple and perfectly fit with actual wind speed data.

In the present work, Weibull distribution is used to fit the wind distribution into different scenarios as defined by Khatod et al. [16]. Accuracy of this approach increases with increase in the number of scenario. Thus six scenarios have been considered in this study to reduce the time span for the calculation. Forecasted 24 h wind speed data have been divided into 24 hly segments and each hourly segment is further divided into $(n + 1)$ discrete states (V_{min}, V_1, V_2 ...V_{n-1}, V_{max}) using Eqs. (8.1) and (8.2). Weibull distribution is a continuous function and different scenarios (S_1, S_2 ..., S_{Ns-1}, S_{Ns}) have been found using these discrete states as given in Eq. (8.4) and shown in Fig. 8.1. The probability density function (PDF) and cumulative distribution function (CDF) in Weibull distribution has been denoted by $f(V)$ and $F(V)$ respectively and are given in Eqs. (8.5) and (8.6).

$$V_1 = V_{min} + \Delta V \tag{8.1}$$

$$V_i = V_{i-1} + \Delta V \tag{8.2}$$

Fig. 8.1 Wind speed probability distribution curve

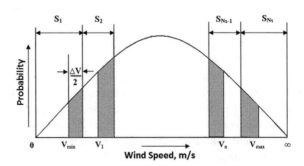

$$\text{where} \quad \Delta V = \frac{V_{\max} - V_{\min}}{n}, \quad V_i = \{V_2, V_3 \ldots V_{\max}\} \tag{8.3}$$

$$S_i = \begin{cases} 0 \to V_{\min} + \frac{\Delta V}{2} & for & i = 1 \\ V_i - \frac{\Delta V}{2} \to V_i + \frac{\Delta V}{2} & for & 1 < i < Ns \\ V_{\max} - \frac{\Delta V}{2} \to \infty & for & i = Ns \end{cases} \tag{8.4}$$

$$f(V) = ba^{-b}V^{b-1}e^{-(V/a)^b} \tag{8.5}$$

$$F(V) = \int_0^\infty ba^{-b}V^{b-1}e^{-(V/a)^b} dt = 1 - e^{-(V/a)^b} \tag{8.6}$$

where, b is the form or shape parameter and a is the scale parameter indicating the main wind speed of the area. Parameters b and a can be obtained by the actual measurement of wind speed at the site. Wind speed V is considered as stochastic variable. PDF across these different scenarios has been calculated using Eqs. (8.5) and (8.6). On the similar lines, distribution function has been calculated for entire 24 segments to develop the wind forecast model. The upper and lower limits for each scenario have been shown in Fig. 8.2a, b respectively.

Based on the wind speed, the probabilistic characteristic of the power generation from wind can be derived. Variable speed wind power generation is being used in this study for better energy capture. The relationship between the input wind speed and output power can be represented approximately by Fig. 8.3. The power output corresponding to the different regions can be represented mathematically by Eq. (8.7).

$$Pw = \begin{cases} 0 & V \le V_{ci} & A & region \\ \phi(V) & V_{ci} < V \le V_r & B & region \\ P_r & V_r < V \le V_{co} & C & region \\ 0 & V > V_{co} & D & region \end{cases} \tag{8.7}$$

8.3   Problem Formulation

Most of the scheduling problems are formulated as optimization problems. These optimization problems easily provide the optimal solution, if the input data are well defined and deterministic. In case input data are uncertain, the probability function is used to describe the data uncertainty. In case, the input data are represented by their corresponding expected values rather than probabilistic value then it may lead to a non-optimal solution. Alternatively, the probability distribution of input data can be represented by sets of scenarios and each set is combined with their probability of occurrence. For such instances, the sum of their probability of occurrence is always equal to one. Thus, by considering input data, the optimization problem can be implemented as a stochastic optimization problem.

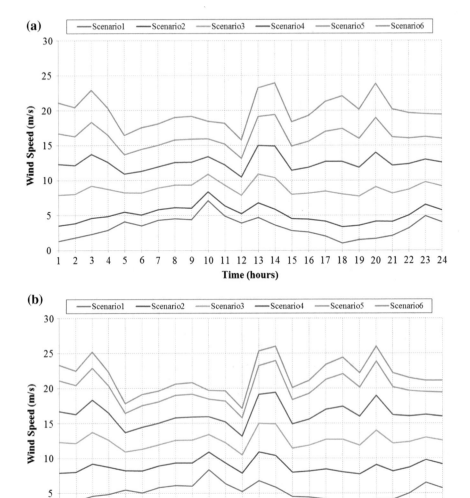

Fig. 8.2 **a** Lower wind speed limit for each Scenario, **b** upper wind speed limit for each scenario

For the present study, stochastic approach has been used to implement the wind-PSP scheduling problem considering the wind system uncertainty as input data. The average sum of each individual solution set of uncertain input data weighted by their associated probability is considered to achieve a single solution. A probability based forecasting model is used to forecast the uncertainty in the input data and probability distribution function has been computed for each set of input data.

Fig. 8.3 Output power
versus wind speed

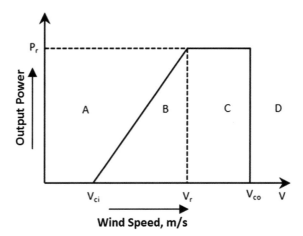

In this study, uncertainty across the wind data is considered in the form of scenarios. Effect of uncertainty on wind-PSP system has been analyzed by maximizing the total revenue (TR) for the given set of scenarios. A stochastic method is adopted for analysis, considering the wind data as a stochastic variable. In this method, probabilistic factor (π) has been assigned in order to compute the TR for the given set of scenarios as shown in Eq. (8.8). Stochastic approach is also known as random approach, which randomly generate the values of the variable within specified scenario limits. So, this stochastic variable is the function of t-th time and s-th scenario. For each scenarios, the value of the stochastic variable for t-th time has been optimized. The final solution for the objective function has been determined by taking algebraic sum of all the stochastic variables weighted by their associated probability Similarly, number of test cases has been considered and each case is defined with different set of scenarios. The Taguchi method is applied (as explained in subsequent section) to select the appropriate combination of scenario based on the result obtained from stochastic method. In this method, orthogonal array structure is used to select the best level of scenarios on the basis of maximum risk.

8.3.1   Wind PSP Model

Wind-PSP model is designed as a stochastic mixed integer type problem, which scheduled their hourly bids for day-ahead market. This scheduling has been done with the objective of "maximizing the hourly profit across both systems" and providing "minimum imbalance across the market". To utilize the power generated by wind and PSP efficiently as per the demand or bid received from the electricity market, a strategy has been developed. In this strategy, both the systems operate as single entity and provide the combined generation to increase the overall revenue. During the low demand period, energy generated by the wind power plant is stored in the upper reservoir by pumping action. This stored energy can be utilized for

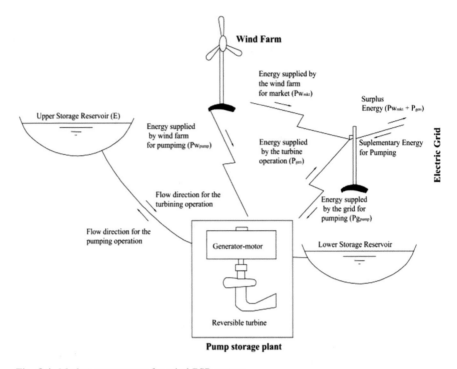

Fig. 8.4 Market arrangement for wind-PSP system

peak period to reduce the market imbalance. A Grid connected wind-PSP system has been shown in Fig. 8.4. In case, wind power plant is unable to supply the power during the pumping mode; the required power for pumping is drawn from the grid.

8.3.2 Objective Function

In the market based system, total revenue or expected profit of the PSP is maximized by operating in generation mode when market price is high and in pumping mode when the price is low. The income of a PSP includes net revenue received by selling energy in the day-ahead market, when it is operating in the generating mode and buying energy during pumping mode. The objective function for this problem has been formulated to maximize the total revenue obtained from the combined operation of pump storage and wind power plant as given in Eq. (8.8).

Maximization of total revenue (TR)

$$\sum_{s=1}^{Ns}\sum_{t=1}^{T}\pi(t,s)\big(R_{psp}(t,s)+R_{wind}(t,s)-R_{loss}(t,s)\big) \tag{8.8}$$

where:

$$R_{psp}(t,s) = R_{gen}(t,s) - R_{pump}(t,s) \tag{8.9}$$

$$R_{gen}(t,s) = s_g(t,s) \times P_{gen}(t,s) \times \lambda_{mkt}(t,s) \tag{8.10}$$

$$R_{pump}(t,s) = Pw_{pump}(t,s) \times \lambda_{wind} + Pg_{pump}(t,s) \times \lambda_{mkt}(t,s) \tag{8.11}$$

$$R_{wind}(t,s) = Pw_{mkt}(t,s) \times \lambda_{mkt}(t,s) \tag{8.12}$$

$$R_{loss}(t,s) = \omega \times \lambda_{mkt}(t) \times [P_d(t) - Pw_{mkt}(t,s) - s(t,s) \times P_{gen}(t,s)] \tag{8.13}$$

In Eq. (8.8), T is the total time period and Ns is the total number of scenarios. The total revenue of PSP at tth hour of sth scenario is given in Eq. (8.9). Revenue from the power generated by PSP $(R_{gen}(t,s))$ is defined as the product of the power generated by PSP $(P_{gen}(t,s))$ during the generating mode and the market price $(\lambda_{mkt}(t,s))$, where $s_g(t,s)$ is the switching variable for generating mode as given in Eq. (8.10). During pumping mode, the PSP uses the power from two sources i.e. one from the wind power plant and other from the grid. The revenue loss during the pumping mode is the sum of revenue loss from the wind power plant generation as well as revenue loss from the grid as given in Eq. (8.11), where λ_{wind} is the price paid by the PSP for utilizing the wind power for pumping operation and remains constant for whole period. $R_{wind}(t,s)$ in Eq. (8.12), is the revenue from the power supplied by wind power plant to the market, referred as $Pw_{mkt}(t,s) \cdot R_{loss}(t,s)$ is the revenue loss or the penalty imposed by the market for creating imbalance between generation and demand as given in Eq. (8.13).

8.3.3   Constraints

The objective function as given in Eq. (8.8) has been formulated subjected to the following equality and inequality constraints:

8.3.3.1   Grid Power for Pumping Constraints

In grid connected day-ahead market, grid power supply to Wind-PSP system for pumping operation is given by Eq. (8.14).

$$Pg_{pump}^{min} \leq Pg_{pump}(t,s) \leq Pg_{pump}^{max} \tag{8.14}$$

8.3.3.2 Wind Power Constraint

The total power generated by the wind power plant is $Pw_{gen}(t, s)$. During the operation, when it is not possible to utilize all the power generated by the wind power plant, some power may remain unutilized. This power is determined by the constraint given in Eq. (8.15).

$$Pw_{unutilized}(t, s) = Pw_{gen}(t, s) - Pw_{pump}(t, s) - Pw_{mkt}(t, s) \qquad (8.15)$$

8.3.3.3 Pumping Constraint

For the pumping operation, PSP can either take the power from the grid or wind power plant as given by the Eq. (8.16).

$$P_p(t, s) \times s_p(t, s) = Pw_{pump}(t, s) + Pg_{pump}(t, s) \qquad (8.16)$$

8.3.3.4 Energy Balance in Upper Reservoir

The energy balance in the upper reservoir is represented by Eq. (8.17). At the beginning of $(t + 1)$th hour, the energy in the upper reservoir $E(t+1, s)$ is the sum of initial level in the reservoir $E(t, s)$ at tth hour, $P_p(t, s) \times [1 - s(t, s)] \times \eta_p$ is the energy supplied to the upper reservoir during pumping mode and $P_{gen}(t, s) \times s_g(t, s)/\eta_h$ is the energy generated from the upper reservoir during generation mode, whereas $s_g(t, s)$ and $s_p(t, s)$ are the mode control variables. t_d is the time deviation or the time interval between each period, η_p and η_h are the efficiency of pumped storage plant during pumping and generating mode respectively.

$$E(t+1, s) = (P_p(t, s) \times s_p(t, s) \times \eta_p - P_{gen}(t, s) \times s_g(t, s)/\eta_h) \times t_d + E(t, s)$$
$$(8.17)$$

8.3.3.5 Upper Reservoir Limit

The energy stored in the upper reservoir has upper and lower limits as given in Eq. (8.18).

$$E^{max} \leq E(t, s) \leq E^{min} \qquad (8.18)$$

8.3.3.6   Generation Limit

This constraint restricts the power generated by the pumped storage plant within the upper and lower limits as given in Eq. (8.19).

$$P_{gen}^{min} \leq P_{gen}(t,s) \leq P_{gen}^{max} \tag{8.19}$$

8.3.3.7   Pumping Limit

This constraint keeps the power consumed by the pumped storage plant within the upper and lower limits and given in Eq. (8.20).

$$P_p^{min} \leq P_p(t,s) \leq P_p^{max} \tag{8.20}$$

8.3.3.8   Switching Constraint

This constraint controls the operation of PSP between the generating and pumping mode and do not allow both operations at the same time as given in Eq. (8.21).

$$s_g(t,s) \times s_p(t,s) = 0 \quad where \quad s_g(t,s), s_p(t,s) \in \{0,1\} \tag{8.21}$$

8.4   Taguchi Method

In the "trial and error" approach, numbers of experiments are performed to observe the variability or uncertainty of parameters, which affect the output results [17]. Main drawback of this approach is its inability to provide optimum result as well as long computation time due to large number of experiments [18]. Wen-Hsien et al. [19] recommended the use of Taguchi method to solve the difficulties in the conventional "trial and error" approach, where, many ideas from the statistical experimental design are used for evaluating and implementing improvements in the process for handling the uncertain parameters.

Taguchi (1987) developed a method for designing experiment to improve the quality of a product by reducing the effect of uncertain variable. The idea used for designing the experiment differs from the other conventional experimental design techniques as it investigates the effect of uncertain parameters by considering the both mean and variance of a process performance characteristics. The basic concept of Taguchi methodology is based on the fractional factorial design [20]. Taguchi method evaluates and implements the improvement in processes, products, facilities and equipment to yield the best result by reducing the number of defects and uncertainties in the design variables. Tsui [21] gave a review of Taguchi's robust design methodology. In the last two decades, Taguchi's method has been applied to a number of design applications and provided the effective results by reducing the number of experiments.

Orthogonal Array (OA) based Taguchi's method provides a simple design for experiments having large number of variables on few levels [22]. The major objectives of the parameter design based techniques are to minimize the product or process variation and to design flexible and robust products or processes that are adaptable to any environmental conditions.

Taguchi method is used in the present study for managing the risk across the wind-PSP system; it is required to find the level of risk across each scenario. One of the major advantages of using Taguchi based risk management strategies across the wind-PSP system is to build the system for bearing the risk during day-ahead market scheduling.

8.5 Methodology

8.5.1 Problem *Description*

It is not necessary that both of the wind and PSP system are installed at same location for the hybridized operation. Many early PSP projects used existing, conventional hydro facilities to provide the necessary upper reservoir for water storage. Modern pumped storage installations are on a larger scale, with most installations having multiple units of 100 MW or greater. So, there are variety of ways so that pumped storage system (PSP) can be implemented within specific geological and hydrological constraints. With proper site selection, the lower reservoir can be an existing water feature such as a river, lake or existing hydro reservoir and it is only necessary to build an upper reservoir. PSP provides peak power and absorbs the surplus/intermittent power. The cost economy of the PSP lies with the differential cost for peaking and off peak power.

The objective function of the study is to maximize the expected profit of the wind and PSP system under the uncertain condition. The total profit represents by taking the sum of the revenue of wind and PSP generation in each scenarios, taking into account the probability of occurrence π. A time horizon of 24 h is considered. Within this time horizon, the wind-PSP system must decide (1) hourly offers to be submitted to day-ahead market, and (2) the system generation in each hour for a given scenario.

8.5.2 Wind *Power Forecasts*

The present research study presents a novel optimal scheduling algorithm for Wind-PSP for different uncertainty scenarios using Taguchi method. Many parameters of wind affect the operation of the Wind-PSP system, out of which wind speed being the most significant in scheduling problem has been identified and predicted.

Forecasts of wind speed are used as input to both the day ahead market participation and the day ahead scheduling modules. In this work, such forecast were obtained using a probabilistic based statistical model, forecasting the wind speed in the form of scenarios as given in the Sect. 8.2. Each scenario has been generated with the length of $(V_{max} - V_{min})/N_S$ using Weibull distribution model, where N_S is the number of scenarios and V_{max} and V_{min} are the maximum and minimum wind speed. Each scenario is assigned with their associated probability as obtained from Weibull distribution model. The relationship between the forecasted wind speed and the output power generation has been provided by using the power curve model as shown in the Fig. 8.3.

8.5.3   Market Imbalance and Price

The day ahead scheduling is based on the minimization of the Revenue loss as given by Eq. (8.8). Revenue loss is the penalty imposed by the market operator for creating imbalance between the actual generation and demand for which generation has to be scheduled by the power operator. The day ahead market, market operator imposed the penalty ω which is considered to be the 75% of the market price for this study. For this study, market price already has been given, whereas the wind price assumed to be remain constant for supplying the power for the pumping operation.

8.5.4   Taguchi's Orthogonal Array Model

For managing the risk across the wind-PSP system, it is required to find the level of risk across each scenario. One of the major advantages of using risk management strategies across the Wind-PSP system is to build the system for bearing the risk during day-ahead market scheduling. An orthogonal array is a matrix that is represented by $L_h(B)^{Ns}$. h and Ns are representing the number of rows and the columns of OA respectively. Total numbers of testing level h in the Taguchi method are decided by orthogonal array (OA), which depends on the value of Ns and B as used in systems [23, 24].

According to the Fraley et al. [25] the proper orthogonal array L_h can be selected using the array selector table as shown in Table 8.1, after knowing the number of variable N_s and the varying level across each variable B. This array selector table was created using an algorithm developed by Taguchi and allowed each variable and setting to be tested equally [26]). For the given Wind-PSP system, $(S_1, S_2, S_3 \ldots S_{Ns})$ are the scenarios representing the wind uncertainty. Each scenario is selected within the two given limits as represented by L (*Low*) and H (*High*) and shown in Fig. 8.2a, b. The algorithm for applying the Taguchi method for Wind-PSP system has been presented below:

(i) Determination of the number of uncertain variables or scenarios and used to predict the actual availability of wind data for the given time period. This data is used to evaluate the total revenue (*TR*) across the Wind-PSP system.

(ii) Each scenario varied between the two limits L (*Low*) and H (*High*). The variation in each scenario effect the value of *TR*.

(iii) Orthogonal arrays indicate the effect of variation in each scenario between the low and high limits. Number of testing in the orthogonal array depends upon the number of scenarios selected (N_s) and the variation limit across each scenario (B). Taguchi method has been applied with a $L_8(2)^6$ Orthogonal array, where the number of testing levels are 8 as shown in Table 8.1.

(iv) After each test, total revenue can be found across Wind-PSP system for each scenario. The main objective of this analysis is to maximize the revenue across each scenario, satisfying the constraints given in Sect. 8.3.3, where (*TR1, TR2…TR8*) are the total revenue of each test level as given in Table 8.2.

(v) To observe the effect of uncertainty in *TR* of the Wind-PSP system, each scenario is divided into two groups. First and second groups are associated with low level and high level respectively and average value (*Avg*) of *TR* across each group has been computed. For example, average value of *TR* across 4th scenario has been calculated as given in Eqs. (8.22) and (8.23), representing the *Avg* for Low and High levels respectively. Variation of each *Avg* value (*dA*) across the respective scenario has been computed to demonstrate the effect of variation of level in Wind-PSP system as shown in Table 8.3.

$$Avg\_L4 = (TR1 + TR3 + TR5 + TR7)/4 \qquad (8.22)$$

$$Avg\_H4 = (TR2 + TR4 + TR6 + TR8)/4 \qquad (8.23)$$

$$dA4 = Avg\_L4 - Avg\_H4 \qquad (8.24)$$

Table 8.1 Orthogonal array selector table

No. of level (B)	No. of variables or scenarios (N_s)								
	2	3	4	5	**6**	7	8	9	10
2	L_4	L_4	L_8	L_8	**L_8**	L_8	L_{12}	L_{12}	L_{12}
3	L_9	L_9	L_9	L_{18}	L_{18}	L_{18}	L_{18}	L_{27}	L_{27}
4	L_{16}	L_{16}	L_{16}	L_{16}	L_{32}	L_{32}	L_{32}	L_{32}	L_{32}
5	L_{25}	L_{25}	L_{25}	L_{25}	L_{25}	L_{50}	L_{50}	L_{50}	L_{50}

Table 8.2 Structure of $L_8(2)^6$ orthogonal array (OA)

No. of testing	Scenarios—uncertain variables						TR
	S_1	S_2	S_3	S_4	S_5	S_6	
1	L1	L2	L3	*L4*	L5	L6	*TR1*
2	L1	L2	L3	**H4**	H5	H6	**TR2**
3	L1	H2	H3	*L4*	L5	H6	*TR3*
4	L1	H2	H3	**H4**	H5	L6	**TR4**
5	H1	L2	H3	*L4*	H5	H6	*TR5*
6	H1	L2	H3	**H4**	L5	H6	**TR6**
7	H1	H2	L3	*L4*	H5	L6	*TR7*
8	H1	H2	L3	**H4**	L5	L6	**TR8**

Table 8.3 Test response of OA

Level	Scenarios—uncertain variables					
	S_1	S_2	S_3	S_4	S_5	S_6
L	Avg_L1	Avg_L2	Avg_L3	*Avg_L4*	Avg_L5	Avg_L6
H	Avg_H1	Avg_H2	Avg_H3	**Avg_H4**	Avg_H5	Avg_H6
dA	dA1	dA2	dA3	dA4	dA5	dA6

8.6 Results and Discussions

For the risk management, it is required to identify the maximum risk available in the system, so that the power producers schedule their output according to the day ahead market. The purpose of this study is to maximize the overall profit of the system under the state with maximum risk and set the operational range of the wind system at which system provides the maximum profit with minimum risk. Such study is very useful in solving optimization problem to manage the risk across the system affected from the single or multiple type of uncertainty and yields the maximum possible profit at minimum risk.

The details of the Wind-PSP system considered in this study has been given in Table 8.4. For this study, PSP is used to provide hedging to a wind system to participate in the day-ahead market. In the PSP system, natural inflow in the upper reservoir is not considered, therefore its working is taken as offline mode. Water stored in upper reservoir is considered in the form of energy stored at the end of PSP operation during each period. Thus, the minimum and maximum capacity of upper reservoir is expressed in the form of energy level. The initial and final energy levels of the upper reservoir are assumed to be equal whereas the effect of energy level in lower reservoir has been neglected. The six sets of scenarios have been considered in Taguchi based orthogonal array structure to generate different test cases for both the case studies presented in Sect. 8.4. Two levels referred as high and low have been selected for each scenario, as given in Table 8.2. For each test

Table 8.4 Wind PSP system

Data for wind farm				
Number of units	Cut in wind speed (m/s)	Cut out wind speed (m/s)	Rated wind speed (m/s)	Rated power (kW per unit)
100	3	26	15	1500

Data for adjustable speed type PSP unit (in MW)			
Mode of Operation	Efficiency %	Maximum	Minimum
Generation	84.4	59.65	0
Pumping	90.0	76.06	46.00

Data for upper reservoir (in MWh)				
Reservoir Type	Maximum	Minimum	Initial	Final
Offline	1500	0	1000	1000

Data for power supplied by grid for pumping		
Maximum (MW)	Minimum (MW)	Total (MWh)
100	0	1500

case, wind energy system is operated separate or combined with PSP to increase the total revenue or profit across the market in case studies I and II respectively. 150 MW wind farm data has been taken from Alta wind energy center (Alta-I), located at California, USA consisting of 100 units of 1.5 MW wind turbines and the 59.65/76.06 MW of pumped storage turbine from Hiwasse Dam unit-2 [27]. For this PSP unit, the size of upper reservoir has been assumed to be 1500 MWh subjected to the condition that its initial and final level should be same, so that the size of reservoir easily reflects to any of practical case. Both wind and pumped storage plant supply the power to the market at the given market price as given by [4]. PSP draws the power from the grid for pumping at given market price whereas pumping used wind power at constant given price of Rs. 548 per MW. Grid based pumping energy cannot be more than 1500 MWh with the maximum power capacity limit of 100 MW as shown in Table 8.6. This has been analyzed, using commercial solver KNITRO for this MINLP type problem.

After optimizing both of the cases for each set of scenario, the profit across all the eight experiment with uncertain variable is presented in Table 8.5. It has been found that with the variation in wind speed for each scenario, the profit varied up to 42% in Case-I and 26% in Case-II. This risk is referred to the loss of the profit with the variation in wind data. The average profit across each scenario for the different level (H and L) of uncertainty is used to compute the risk in the form of "loss in profit" as shown in Tables 8.6 and 8.7. The value of risk is the difference of average profit across the level H and L and this value indicates the influence of each factor determined by experiments. The value of risk presented in Tables 8.6 and 8.7, indicated the reduction in risk by negative (−ve) values and increase in the risk by positive values (+ve). For the Case-I, the value of risk obtained from scenario S_1, S_2, S_3, S_4, S_5, S_6 are −77,811.3, −359,259, −185,467, −12,797.3, −839.75, −0.25 respectively as shown in Fig. 8.5. From the Table 8.6, it is seen that the maximum

Table 8.5 Orthogonal array with test result

No. of testing	Scenarios—uncertain variables						Case-I	Case-II
	S_1	S_2	S_3	S_4	S_5	S_6	TR (in Rs.)	TR (in Rs.)
1	L1	L2	L3	L4	L5	L6	767,638	1,321,597
2	L1	L2	L3	H4	H5	H6	781,275	1,325,319
3	L1	H2	H3	L4	L5	H6	1,312,364	1,779,438
4	L1	H2	H3	H4	H5	L6	1,326,001	1,783,159
5	H1	L2	H3	L4	H5	L6	1,031,756	1,528,224
6	H1	L2	H3	H4	L5	H6	1,043,714	1,531,972
7	H1	H2	L3	L4	H5	H6	1,205,548	1,728,845
8	H1	H2	L3	H4	L5	L6	1,217,505	1,731,544

Table 8.6 Revenue across each scenario for case-I

Level	Revenue across each scenario (in Rs.)					
	S_1	S_2	S_3	S_4	S_5	S_6
L	1,046,820	906,096	992,992	1,079,327	1,085,305	1,085,725
H	1,124,631	1,265,355	1,178,459	1,092,124	1,086,145	1,085,725
Risk or Profit Loss (L − H)	−77,811.3	−359,259	−185,467	−12,797.3	−839.75	−0.25
Rank	4	6	5	3	2	1
Optimal set	L1	L2	L3	L4	L5	L6

Table 8.7 Revenue across each scenario Case-II

Level	Revenue across each scenario (in Rs.)					
	S_1	S_2	S_3	S_4	S_5	S_6
L	1,547,195	1,421,557	1,521,871	1,589,464	1,591,144	1,591,297
H	1,630,214	1,755,851	1,655,538	1,587,944	1,586,265	1,586,111
Risk or profit Loss (L − H)	−83,019	−334,294	−133,667	1520	4879	5186
Rank	4	6	5	3	2	1
Optimal set	L1	L2	L3	H4	H5	H6

risk has been with scenario S_1, S_2, S_3, S_4, S_5 and S_6 by setting their level to L. The set of scenarios for case-I, which attain maximum risks are mentioned as L1, L2, L3, L4, L5 and L6. Figure 8.7 shows the generation scheduling of alone wind system for these scenarios. For the Case-II, the negative −ve values of risk obtained from scenario S_1, S_2, S_3 (e.g. −83,019, −334,294, −133,667 respectively) and the positive (+ve) values of risk obtained from the scenario S_4, S_5, S_6 (1520, 4879, 5186 respectively) as shown in Fig. 8.6. From the Table 8.7, it is seen that the maximum risk has been with scenario S_1, S_2, S_3 by setting their level to L, while setting S_4, S_5, S_6 to H. The set of scenarios for case-II, which attain maximum risks are mentioned as L1, L2, L3, H4, H5 and H6. Figures 8.8 and 8.9 show the results of Wind-PSP operation for these scenarios.

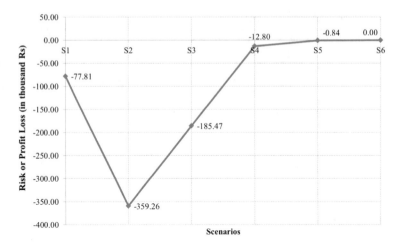

Fig. 8.5 Risk or profit loss across each scenario (case-I)

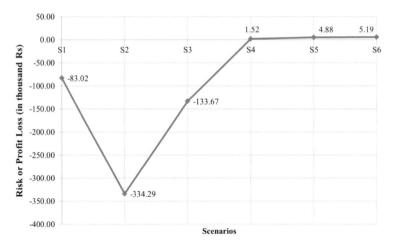

Fig. 8.6 Risk or profit loss across each scenario (case-II)

From Figs. 8.8 and 8.9, it is seen that PSP units are scheduled to generate power during the periods of high market price. It is also observed that, scheduling reduced the imbalance between the generation and demand, thus avoided penalty during high market price for difference in demand and supply. During 1st, 2nd and 3rd scenario, wind power output is very low causing high market imbalance. However, during 4th, 5th and 6th scenario, wind power output is very high, thus market imbalance becomes very low as shown in Table 8.9.

The result comparison for both of the cases has been shown in Tables 8.5, 8.6, 8.7, 8.8 and 8.9. The effect of uncertainty has been considered in both of the cases. It has been seen that the result provided by case-I having high amount of risk due to

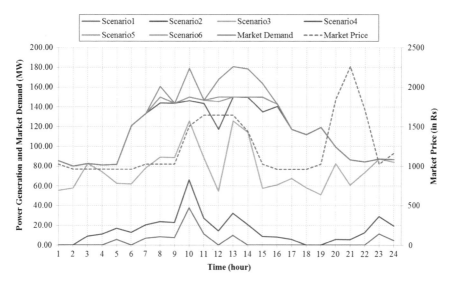

Fig. 8.7 Total power generation and market demand (case-I)

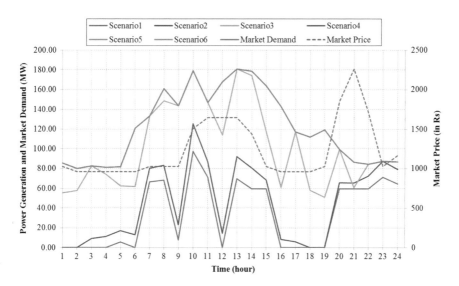

Fig. 8.8 Total power generation and market demand (case-II)

alone operation of Wind system. This risk has been reduced in case-II by providing the combined operation with PSP. Table 8.8 presents that the expected profit of combined wind-PSP system has been increased by nearly 42% as compared to the alone wind system operation. The combined wind-PSP operation does not only increase the profit but also increased the total wind utilization across the system as shown in Table 8.9.

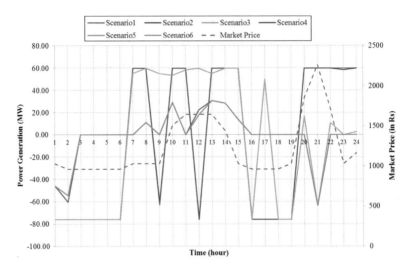

Fig. 8.9 PSP operation with change in market price (case-II)

Table 8.8 Expected revenue across wind-PSP system

Cases	Total expected profit (in thousand Rs.)
With wind system only	767.64
With wind-PSP system	1325.32

Table 8.9 Unutilized wind power and market imbalance

Scenario	Unutilized wind		Market imbalance	
	Case-I	Case-II	Case-I	Case-II
S_1	0.0	0.0	2818.04	2102.24
S_2	0.0	0.0	2548.42	1834.05
S_3	14.69	12.36	1079.86	508.78
S_4	673.48	463.89	194.34	0.0
S_5	803.54	633.37	135.39	0.0
S_6	807.60	643.30	130.70	0.0

The amount of pumping power supplied by the wind and grid system has been shown in Table 8.10, which illustrates that nearly 15–30% of unutilized wind power in case-I, has been successfully utilized by the PSP during the case-II. Some time, wind system is not being able to supply the sufficient power for the pumping operation; in that case remaining power for pumping has been utilized from the grid. As shown in Table 8.10, during the scenario S_1 and S_2, mostly all the pumping power has been utilized from the grid due to unavailability of wind power at that time, whereas large quantity of wind power is available during the scenario S_5 and S_6 and no power has been utilized from the grid for the pumping operation.

Table 8.10 Power supplied for pumping in case-II

Scenario	Power supplied for pumping (in MW)		Total pumping (in MW)
	Wind	Grid	
S_1	0.0	899.70	899.70
S_2	0.0	897.90	897.90
S_3	2.33	715.47	717.80
S_4	209.58	34.69	244.27
S_5	170.17	0.0	170.17
S_6	164.30	0.0	164.30

8.7   Conclusions

In the present study, stochastic mixed integer type problem has been formulated with the objective of maximizing the profit in the wind-PSP system by taking into account the uncertainty in the wind data. Weibull distribution based forecasting model has been used for short term scheduling to forecast the wind data uncertainty. The combined operation of wind and PSP was analyzed to reduce the market imbalance between demand and generation. Uncertainty in the wind not only affects the market imbalance but also reduces the total revenue or profit. Uncertainties in wind speed have been analyzed by Taguchi method. This method evaluates the various level of risk, across wind-PSP system by partial set of experiments and offers a simple, systematic approach. This method used orthogonal array structure to reduce number of experiment to save the computation time and provide the efficient operation of wind-PSP system under uncertain conditions.

In the proposed model for the scheduling of wind-PSP system under day ahead market, the effect of only wind data uncertainty has been considered, however, in practical case, the operation of the power system is also effected by the uncertainty in the market price and demand. These factors should be considered in the future studies. Future studies can be based on Game theory or multi objective based techniques to consider the two or more uncertainty effect of the market. The wear and tear cost and the efficiency of the PSP unit is affected by the number of startups and shutdowns during the operation. The frequent switching of the unit not only reduce its useful life by 10–15 h for each startup or shutdown but also result in loss of energy in the form of water. It is required to minimize the PSP unit cycling with minimum startup and shutdown of the unit to decrease the wear and tear cost and the energy loss. These all factors should be considered in the future research work.

References

1. Li, X., & Jiang, C. (2011). Short term operation model and risk management for wind power penetrated system in electricity market. *IEEE Transactions on Power Systems, 26*(2), 932–939.
2. Li, X., & Jiang, C. (2010). Unit commitment and risk management based on wind power penetrated system. In *IEEE international conference on power system technology*, Hangzhou, October 24–29, 2010.
3. Abreu, L. V. L., Khodayar, M. E., Shahidehpour, M., & Wu, L. (2012). Risk constrained coordination of cascaded hydro unit with variable wind power generation. *IEEE Transaction on Sustainable Energy, 3*(3), 359–368.
4. Catalao, J. P. S., Pousinho, H. M. I., & Mendes, V. M. F. (2012). Optimal offering strategies for wind power producer considering uncertainty and risk. *IEEE Systems Journal, 6*(2), 270–277.
5. Gimeno-Gutiérrez, M., & Lacal-Arántegui, R. (2013). Assessment of the European potential for pumped hydropower energy storage. *JRC scientific and policy reports, European commission.* http://ec.europa.eu/dgs/jrc/downloads/jrc_20130503_assessment_european_phs_potential.pdf.
6. Miller, R. (2011). Renewable energy goals and pumped storage hydropower. http://livebettermagazine.com/article/renewable-energy-goals-pumped-storage-hydropower.
7. Rognlien, L. M. (2012). *Pumped storage development in Øvre Otra, Norway.* Norwegian University of Science and Technology (NTNU). http://www.diva-portal.org/smash/get/diva2:566196/FULLTEXT01.pdf.
8. IRENA. (2012). *Electricity storage and renewables for island power.* International Renewable Energy Agency. https://www.irena.org/DocumentDownloads/Publications/ElectricityStorageandREforIslandPower.pdf.
9. Lu, M.-S., Chang, C.-L., Lee, W.-J., & Wang, L. (2009). Combining the wind power generation system with energy storage equipment. *IEEE Transactions on Industry Applications, 45*(6), 2109–2115.
10. Muyeen, S. M., Takahashi, R., Murata, T., & Tamura, J. (2009). Integration of an energy capacitor system with a variable-speed wind generator. *IEEE Transactions on Energy Conversion, 24*(3), 740–749.
11. Strbac, G., et al. (2012). *Strategic assessment of the role and value of energy storage systems in the UK low carbon energy future.* Energy Futures Lab, Imperial College London. https://www.carbontrust.com/media/129310/energy-storage-systems-role-value-strategic-assessment.pdf.
12. Yao, D. L., Choi, S. S., Tseng, K. J., Lie, T. T. (2009). A statistical approach to design of a dispatchable wind power-battery energy storage system. *IEEE Transactions on Energy Conversion, 24*(4), 961–925.
13. Caralis, G., Papantonis, D., & Zervos, A. (2012). The role of pumped storage systems towards the large scale wind integration in the Greek power supply system. *Renewable and Sustainable Energy Reviews, 16*(5), 2558–2565.
14. Garcia-Gonzalez, J., de la Muela, R. M. R., Santos, L. M., & Gonzalez, A. M. (2008). Stochastic joint optimization of wind generation and pumped-storage units in an electricity market. *IEEE Transactions on Power System, 23*(2), 460–468.
15. Angarita, J. L., Usaola, J., & Martínez-Crespo, J. (2009). Combined hydro-wind generation bids in a pool-based electricity market. *Electric Power Systems Research, 79*(7), 1038–1046.
16. Khatod, D. K., Pant, V., & Sharma, J. (2010). Analytical approach for well-being assessment of small autonomous power system with solar and wind energy sources. *IEEE Transactions on Energy Conversation, 25*(2), 535–545.
17. Sayuti, A., Sarhan, A. A. D., Fadzil, M. & Hamdi, M. (2012). Enhancement and verification of a machined surface quality for glass milling operation using CBN grinding tool—Taguchi approach. *International Journal of Advanced Manufacture Technology, 60*(9–12), 969–950.

18. Wei-Chung, W., Yang, F., & Elsherbeni, A. Z. (2007). Linear antenna array synthesis using Taguchi's method: A novel optimization technique in Electromagnetics. *IEEE Transactions on Antennas and Propagation, 55*(3), 723–730.
19. Wen-Hsien, H., Jinn-Tsong, T., Gong-Ming, H., & Jyh-Horng, C. (2010). Process Parameters Optimization: A Design Study for TiO2 Thin Film of Vacuum Sputtering Process. *IEEE Transactions on Automation Science and Engineering, 7*(1), 143–146.
20. Montgomery, D. C. (1997). *Design and Analysis of Experiments* (4th ed., p. 720). John Wiley & Sons, Canada, ISBN: 0-471-15746-5.
21. Tsui, K. L. (1992). An overview of Taguchi method and newly developed statistical methods for robust design. *IIE Transactions, 24*, 44–57.
22. Sibalija, T. V., Majstorovic, V. D., & Miljkovic, Z. D. (2011). An intelligent approach to robust multiresponse process design. *International Journal of Production Research, Article in Press, 49*(17), 5079–5097.
23. Liu, D., & Cai, Y. (2005). Taguchi method for solving the economic dispatch problem with non-smooth cost functions. *IEEE Transactions on Power Systems, 20*(4), 2006–2014.
24. Yu, H., & Rosehart, W. D. (2012). An optimal power flow algorithm to achieve robust operation considering load and renewable generation uncertainties. *IEEE Transaction on Power Systems, 27*(4), 1808–1817.
25. Fraley, S., Oom, M., Terrien, B., & Zalewski, J. (2007). Design of experiments via Taguchi methods: Orthogonal arrays. *University of Michigan Chemical Engineering Process Dynamics and Controls Open Textbook.* https://controls.engin.umich.edu/wiki/index.php/Design_of_experiments_via_taguchi_methods:_orthogonal_arrays.
26. Ross, P. J. (1995). *Taguchi techniques for quality engineering. Loss function, orthogonal experiments, parameter and tolerance design* (2nd ed., p. 329). McGraw Hill Publishing Co.
27. Hiwassee Dam Unit 2 Reversible Pump-Turbine. (1956). A national historic mechanical engineering landmark. *The American society of Mechanical Engineers (ASME).* Accessed on Oct 29, 2013. http://files.asme.org/asmeorg/Communities/History/Landmarks/5567.pdf.

Index

© Springer Nature Singapore Pte Ltd. 2018
M. Majumder (ed.), *Application of Geographical Information Systems and Soft Computation Techniques in Water and Water Based Renewable Energy Problems*, Water Resources Development and Management, https://doi.org/10.1007/978-981-10-6205-6

Printed in the United States
By Bookmasters